# Algebra 1 Help

By Kathryn Paulk
Copyright © 2024

Updated: 09/10/2024

# Table of Contents

Introduction .................................................. 1

*Algebra Topics - Part 0* ............................... 3

   The Number System ............................... 4

   Exponents ................................................ 11

   Math Operations ..................................... 15

   Fractions ................................................. 18

   Factoring ................................................. 24

   Cartesian Coordinate System ................ 28

   Expressions, Equations, Inequalities ..... 34

   Functions ................................................ 40

   Linear and Quadratic Functions ............ 44

   Parent Functions ................................... 53

   Piecewise Functions .............................. 55

   Set Notation .......................................... 57

   Domain and Range ................................ 59

   More Linear Equations ......................... 62

   Absolute Value Problems ..................... 72

Scientific Notation ............................................83
Making Choices – Bike Rental......................88

## *Algebra Topics - Part 1* ...............................98
Polynomial Division (Long & Synthetic) ........99
Factoring Polynomials ................................105
Why Factor Polynomials.............................116
Completing The Square ..............................118
Quadratic Formula .....................................123
Function Composition ................................133
Inverse Functions .......................................138
Transformations.........................................144
Graphing Quadratic Functions ....................160
Graphing Polynomial Functions..................172
Graphing Rational Functions ......................179
Working With Radicals ...............................195
Logarithms .................................................208
Exponential Growth and Decay ..................223
Regression ..................................................230

Systems of Linear Equations ........................ 247

Systems of Linear Inequalities .................... 253

Solving Inequalities ..................................... 257

*Trig Topics - Part 1* .............................................. 265

Right Triangles ............................................ 266

The Unit Circle ............................................. 273

Trigonometric Functions ............................ 282

Trig Functions of General Angles ................ 291

Law of Cosines ............................................ 300

Law of Sines ................................................ 304

General Triangles ........................................ 309

Bearings ...................................................... 321

*References* .......................................................... 325

*Other Books by Kathryn Paulk* ........................ 327

## **Introduction**

This book will help students who are currently taking or planning to take a course in Algebra 1. For each topic, key equations are listed and followed by detailed examples.

This book has been formatted so that it is easy to read on both paperback and also on electronic devices with the Kindle app (laptop, iPad, Kindle E-reader, and iPhone).

Please note that four "Help" books with different titles contain overlapping content.

| Book Title | Book Content | | | | | |
|---|---|---|---|---|---|---|
| | Alg. 0 | Alg. 1 | Alg. 2 | Trig. 1 | Trig. 2 | Trig. 3 |
| Algebra 1 Help | X | X | | X | | |
| Algebra 2 Help | | X | X | X | X | X |
| Pre-Calc. & Trig. Help | | X | X | X | X | X |
| College Algebra Help | X | X | X | X | X | |

# **Algebra Topics - Part 0**

# The Number System

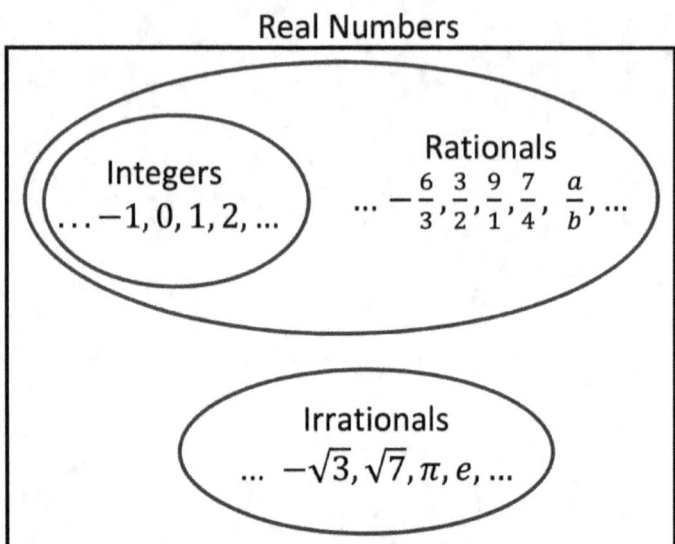

Imaginary Numbers

$a + bi \quad ; b \neq 0, \quad i = \sqrt{-1}$

### Natural Numbers

Natural numbers are counting numbers. They were used to count things like sheep, potatoes, and other physical items. Natural numbers are: 1, 2, 3, ...
Natural numbers do not include negative numbers or zero.

### Whole Numbers

Whole numbers include zero with the set of natural numbers. The concept of zero was a very advanced idea. WhOle numbers are: **0**, 1, 2, 3, ...
Whole numbers do not include negative numbers.

### Integers

Integers include negative and positive whole numbers.
Integers: $\quad$ ... $-3, -2, -1, 0, 1, 2, 3,$
Negative Integers: ... $-3, -2, -1$
Positive Integers: $\quad 1, 2, 3, ...$
Zero is not included in negative or positive integers.
But... Zero is included in Integers.

## Rational Numbers

Rational numbers can be written in the form: $\frac{a}{b}$

Where $a$ and $b$ are integers.

Some rational numbers are also integers. $\left(e.g. \ \frac{6}{2}\right)$

Rational numbers may also be written in decimal form. For example, $\frac{3}{2}$ can be written as $1.5$.

$\frac{1}{3}$ can be written as $1.33\overline{3}$

If the decimal terminates or repeats, then it can be written as a fraction, so it is a rational number.

If a decimal number does not terminate or repeat, then it cannot be written a fraction, so it is an irrational number.

## Irrational Numbers

Irrational numbers cannot be written as a fraction, $\frac{a}{b}$

Where $a$ and $b$ are integers. In decimal form, irrational numbers do not terminate or repeat.

For example, $\sqrt{7} = 2.6457\ldots$ and $\pi = 3.14159\ldots$

| Real Numbers |
|---|
| Real numbers include integers, rational numbers, and irrational numbers. |

| Imaginary Numbers |
|---|
| Imaginary numbers are in the form: $a + bi$ <br> Where $b \neq 0$ <br> Therefore, all imaginary numbers have an imaginary component. The real part ($a$) may be zero. |
| By definition: $i = \sqrt{-1}$ <br> Therefore: <br> $\quad i^2 = -1$ <br> $\quad i^3 = (i^2)^1 \cdot i \quad = (-1) \cdot i \quad = -i$ <br> $\quad i^4 = (i^2)^2 \quad\quad = (-1)^2 \quad\quad = 1$ <br> $\quad i^5 = (i^2)^2 \cdot i \quad = (-1)^2 \cdot i \quad = i$ <br> $\quad i^6 = (i^2)^3 \quad\quad = (-1)^3 \quad\quad = -1$ <br> $\quad i^7 = (i^2)^3 \cdot i \quad = (-1)^3 \cdot i \quad = -i$ <br> $\quad i^8 = (i^2)^4 \quad\quad = (-1)^4 \quad\quad = 1$ <br><br> Also: $\sqrt{-9} = \sqrt{(9)(-1)} = \sqrt{9} \cdot \sqrt{-1} = 3i$ |

## Rational Numbers
### Convert Fractions to Decimals -- Examples

| | |
|---|---|
| Convert $\frac{5}{4}$ to a decimal  Answer: 1.25 | ``` 1. 2 5 4 ) 5. 0 0 4 ─── 1 0 0 8 ─── 2 0 2 0 ─── 0 ``` |
| Convert $\frac{7}{3}$ to a decimal  Answer: $2.\overline{3}$ | ``` 2. 3 3 3 3 ) 7. 0 0 0 6 ─── 1 0 0 9 ─── 1 0 0 9 ─── 1 0 ``` |

Note: A bar over a number means it repeats.

| Rational Numbers | |
|---|---|
| **Convert Terminating Decimal to Fraction -- Examples** | |
| Convert 1.2 to a fraction | $1.2 = \frac{12}{10} = \frac{6}{5}$ |
| Convert 1.25 to a fraction | $1.25 = \frac{125}{100} = \frac{5}{4}$ |

| Rational Numbers | |
|---|---|
| **Convert Repeating Decimal to Fraction -- Examples** | |
| Convert $1.\overline{23}$ to a fraction | $\begin{aligned} 100x &= 123.\overline{23} \\ -\quad x &= \phantom{00}1.\overline{23} \\ \hline 99x &= 122 \end{aligned}$ <br> Solve for $x$: $x = \frac{122}{99}$ |
| Convert $12.\overline{345}$ to a fraction | $\begin{aligned} 1000x &= 12345.\overline{345} \\ -\quad x &= \phantom{000}12.\overline{345} \\ \hline 999x &= 12333 \end{aligned}$ <br> Solve for $x$: $x = \frac{12333}{999} = \frac{4111}{333}$ |

## The Number System
### Classify the Given Numbers -- Examples

| Number | $-3$ | 3.1 | $3.\overline{12}$ | 5.126 ... | $\sqrt{-4}$ |
|---|---|---|---|---|---|
| Integer | X | | | | |
| Rational | X | X | X | | |
| Irrational | | | | X | |
| Real | X | X | X | X | |
| Imaginary | | | | | X |

# Exponents

| Exponents -- Rules | |
|---|---|
| $a^0$ | $= 1$ |
| $a^1$ | $= a$ |
| $a^2$ | $= a \cdot a$ |
| $a^m \cdot a^n$ | $= a^{m+n}$ |
| $\dfrac{a^m}{a^n}$ | $= a^{m-n}$ |
| $\dfrac{a^n}{a^n}$ | $= a^{n-n} = a^0 = 1$ |
| $(a \cdot b \cdot c)^n$ | $= a^n \cdot b^n \cdot c^n$ |
| $(a^m)^n$ | $= a^{m \cdot n}$ |
| $a^{-n}$ | $= \dfrac{1}{a^n}$ |
| $\left(\dfrac{a}{b}\right)^{-n}$ | $= \left(\dfrac{b}{a}\right)^n = \dfrac{b^n}{a^n}$ |
| $\sqrt[n]{a^m}$ | $= a^{\left(\frac{m}{n}\right)}$ |
| $\sqrt[n]{a^n}$ | $= a^{\frac{n}{n}} = a^1 = a \quad$ if n is odd <br> $= a^{\frac{n}{n}} = \lvert a \rvert \quad$ if n is even |

## Exponents – Examples 01

Simplify the following expressions.

| Expression | Simplified |
|---|---|
| $128^0$ | $= 1$ |
| $456^1$ | $= 456$ |
| $3^2$ | $= 3 \cdot 3 = 9$ |
| $3^2 \cdot 3^4$ | $= 3^{2+4} = 3^6$ <br> $= 3 \cdot 3 \cdot 3 \cdot 3 \cdot 3 \cdot 3 = 729$ |
| $\dfrac{5^{10}}{5^4}$ | $= 5^{10-4} = 5^6 = 15625$ |
| $(2xy)^4$ | $= 2^4 \cdot x^4 \cdot y^4 = 16x^4y^4$ |
| $(2^3)^5$ | $= 2^{3 \cdot 5} = 2^{15} = 32768$ |
| $5^{-2}$ | $= \dfrac{1}{5^2} = \dfrac{1}{25}$ |
| $\left(\dfrac{1}{4}\right)^{-2}$ | $= \left(\dfrac{4}{1}\right)^2 = \dfrac{4^2}{1^2} = \dfrac{16}{1} = 16$ |
| $\sqrt[4]{9^2}$ | $= 9^{\left(\frac{2}{4}\right)} = 9^{\frac{1}{2}} = \sqrt[2]{9} = \sqrt{9} = 3$ |
| $25^{\frac{1}{2}}$ | $= \sqrt[2]{25^1} = \sqrt{25} = 5$ |

## Exponents – Examples 02

Simplify the following expressions.

| Expression | Simplified |
|---|---|
| $\dfrac{10\,a^4\,b^2}{5\,a^2\,b^8}$ | $= \dfrac{10}{5} \cdot \dfrac{a^4}{a^2} \cdot \dfrac{b^2}{b^8} = \dfrac{5\,a^2}{b^6}$ |
| $(3^2 x^5 y)^4$ | $= 3^8 \cdot x^{20} \cdot y^4$ |
| $(2^3)^x$ | $= 2^{3x}$ |
| $(-5)^{-2}$ | $= \dfrac{1}{(-5)^2} = \dfrac{1}{25}$ |
| $-\left(\dfrac{1}{4}\right)^{-2}$ | $= -\left(\dfrac{4}{1}\right)^2 = -\dfrac{4^2}{1^2} = -16$ |
| $\sqrt{18}$ | $= \sqrt{9 \cdot 2} = \sqrt{9} \cdot \sqrt{2} = 3\sqrt{2}$ |
| $\sqrt{-9}$ | $= \sqrt{9\,(-1)} = \sqrt{9} \cdot \sqrt{-1} = 3i$ |
| $\sqrt{-32}$ | $= \sqrt{16 \cdot 2 \cdot (-1)} = 4\sqrt{2}\,i$ |
| $\sqrt[3]{8}$ | $= 2 \qquad$ Because $2 \cdot 2 \cdot 2 = 8$ |
| $\sqrt[3]{-8}$ | $= -2 \qquad$ Because $(-2)^3 = -8$ |
| $\sqrt{9x^2}$ | $= 3|x|$ |

# Math Operations

## Math Operations

Math terms are multiplied numbers or variables.
Some examples:  2, $2x$, $5x^2$, $3\sqrt{2}$, $\frac{5x}{2}$, $4abc$

Math operations are applied to math terms.
Some basic math operations are:
- Exponents
- Multiplication and division
- Addition and subtraction.

## Order of Operations

Sometimes math terms and operations may be grouped together with parenthesis. Always evaluate the terms within the parenthesis first. Then, apply the math operations in the order, listed above.
Some students find the acronym PEMDAS helpful.
- P = Parenthesis
- E = Exponents
- M D = Multiplication & Division
- A S = Addition & Subtraction

A simpler way to think about it is: Do the stuff in the parenthesis first. Then, do the hard stuff first!

## Math Operations -- Examples 01

Simplify the math expressions by applying the math operations in the correct order.

| Expression | Simplified Expression |
|---|---|
| $3 + (5 - 7)$ | $3 + (5 - 7)$ <br> $3 + (-2)$ <br> $3 - 2$ <br> $1$ |
| $3 - 2(8 - 3)$ | $3 - 2(8 - 3)$ <br> $3 - 2(5)$ <br> $3 - 10$ <br> $-7$ |
| $10 - 3(8 - 6)^2$ | $10 - 3(8 - 6)^2$ <br> $10 - 3(2)^2$ <br> $10 - 3 \cdot 4$ <br> $10 - 12$ <br> $-2$ |
| $12 - (3 - x)^2$ | $12 - (3 - x)^2$ <br> $12 - (3 - x)(3 - x)$ <br> $12 - [9 - 3x - 3x + x^2]$ <br> $12 - [9 - 6x + x^2]$ <br> $12 - 9 + 6x - x^2$ <br> $3 + 6x - x^2$ |

# Fractions

## Fractions

Fractions are rational numbers in the form: $\frac{a}{b}$
Where: $a = numerator$. $b = denominator \neq 0$

It may be helpful to think of fractions as pieces of pizza. If you cut a pizza into 8 pieces, each piece is 1/8 of the pizza.

| | |
|---|---|
| Addition and Subtraction | When adding or subtracting fractions, the denominators must be the same. |
| | $\frac{5}{8} + \frac{3}{8} = \frac{8}{8} = 1$ (one pizza) |
| | $\frac{5}{8} - \frac{3}{8} = \frac{2}{8} = \frac{1}{4}$ |
| Multiplication | When multiplying fractions, just multiply straight across. |
| | $\frac{2}{3} \cdot \frac{5}{7} = \frac{10}{21}$ |
| Division | When dividing by a fraction, invert it (flip it) and then multiply. |
| | $\frac{2}{3} \div \frac{5}{7} = ???$ |
| | $\frac{2}{3} \cdot \frac{7}{5} = \frac{14}{15}$ |

| Fractions -- More ||
|---|---|
| Another way to think about fractions is to visualize cookies on a table. | |
| If there are 4 cookies on the table and two people, seated at the table. Then, each person gets 2 cookies. | $\frac{4}{2} = 2$ |
| If there are 6 cookies on the table and two people, seated at the table. Then, each person gets 3 cookies. | $\frac{6}{2} = 3$ |
| If there are 0 cookies on the table and two people, seated at the table. Then, each person gets 0 cookies. | $\frac{0}{2} = 0$ |

Important Notes:

- Zero divided by anything is zero. $\frac{0}{n} = 0$
- Anything divided by zero is undefined. $\frac{n}{0} = \infty$

## Fractions -- Addition & Subtraction Examples

Multiplying any number by 1 does not change it. To change a denominator, it is often helpful to multiply by 1 in the form $\frac{n}{n}$. Note: $\frac{n}{n} = 1$

| Expression | Simplified Expression |
|---|---|
| $\frac{1}{2} + \frac{1}{4}$ | $\frac{1}{2}\left(\frac{2}{2}\right) + \frac{1}{4}$ <br> $\frac{2}{4} + \frac{1}{4} = \frac{3}{4}$ |
| $\frac{1}{2} - \frac{3}{7}$ | $\frac{1}{2}\left(\frac{7}{7}\right) - \frac{3}{7}\left(\frac{2}{2}\right)$ <br> $\frac{7}{14} - \frac{6}{14} = \frac{1}{14}$ |
| $\frac{1}{a} + \frac{2}{b}$ | $\frac{1}{a}\left(\frac{b}{b}\right) + \frac{2}{b}\left(\frac{a}{a}\right)$ <br> $\frac{b}{ab} + \frac{2a}{ab} = \frac{b+2a}{ab}$ |
| $\frac{1}{a^2} + \frac{2}{b} - \frac{3}{ac}$ <br><br> Hint: Smallest common denominator is $a^2bc$ | $\frac{1}{a^2}\left(\frac{bc}{bc}\right) + \frac{2}{b}\left(\frac{a^2c}{a^2c}\right) - \frac{3}{ac}\left(\frac{ab}{ab}\right)$ <br> $\frac{bc}{a^2bc} + \frac{2a^2c}{a^2bc} - \frac{3ab}{a^2bc}$ <br> $\frac{bc + 2a^2c - 3ab}{a^2bc}$ |

## Fractions -- Multiplication Examples

When multiplying fractions, just multiply straight across.

When possible, simplify fractions by dividing the numerator and denominator by the same number.

| Expression | Simplified Expression |
|---|---|
| $\dfrac{1}{2} \times \dfrac{1}{4}$ | $\dfrac{1}{2} \times \dfrac{1}{4} = \dfrac{1}{8}$ |
| $\dfrac{3}{4} \times \dfrac{100}{6}$ | $\dfrac{1}{2} \times \dfrac{100}{3} = \dfrac{100}{6} = \dfrac{50}{3}$ |
| $\dfrac{1}{a} \times \dfrac{2}{b}$ | $\dfrac{1}{a} \times \dfrac{2}{b} = \dfrac{2}{ab}$ |
| $\dfrac{1}{a^2} \times \dfrac{2}{b} \times \dfrac{3}{ac}$ | $\dfrac{1}{a^2} \times \dfrac{2}{b} \times \dfrac{3}{ac} = \dfrac{6}{a^3 bc}$ |
| $\dfrac{b}{a^2} \times \dfrac{c}{b} \times \dfrac{3}{ac}$ | $\dfrac{b}{a^2} \times \dfrac{b}{b} \times \dfrac{3}{ac} = \dfrac{3bc}{a^3 bc} = \dfrac{3}{a^3}$ |
| $\dfrac{1}{2} \times 66$ | $\dfrac{1}{2} \times 66 = \dfrac{1}{2} \times \dfrac{66}{1} = \dfrac{66}{2} = 33$ |
| $\dfrac{1}{3} \times 66$ | $\dfrac{1}{3} \times 66 = \dfrac{1}{3} \times \dfrac{66}{1} = \dfrac{66}{3} = 22$ |
| $\dfrac{2}{2} \times 66$ | $\dfrac{2}{2} \times 66 = \dfrac{2}{2} \times \dfrac{66}{1} = \dfrac{132}{2} = 66$ |

## Fractions -- Division Examples

Dividing by a fraction may be expressed in two ways:

$$\frac{a}{b} \div \frac{c}{d} = \frac{a}{b} \times \frac{d}{c} = \frac{ad}{bc}$$

Or

$$\frac{\left(\frac{a}{b}\right)}{\left(\frac{c}{d}\right)} = \frac{a}{b} \times \frac{d}{c} = \frac{ad}{bc}$$

| Expression | Simplified Expression |
|---|---|
| $\frac{1}{2} \div \frac{1}{4}$ | $= \frac{1}{2} \times \frac{4}{1} = \frac{4}{2} = \frac{2}{1} = 2$ |
| $\frac{3}{4} \div \frac{100}{6}$ | $= \frac{3}{4} \times \frac{6}{100} = \frac{18}{400} = \frac{9}{200}$ |
| $\frac{1}{a} \div \frac{2}{b}$ | $= \frac{1}{a} \times \frac{b}{2} = \frac{b}{2a}$ |
| $\frac{\left(\frac{1}{a}\right)}{\left(\frac{2}{b}\right)}$ | $= \frac{1}{a} \times \frac{b}{2} = \frac{b}{2a}$ |
| $\frac{3}{4} \div \frac{(x+2)}{6}$ | $= \frac{3}{4} \times \frac{6}{(x+2)} = \frac{18}{4(x+2)} = \frac{9}{2(x+2)}$ |
| $\frac{3}{(x+2)} \div \frac{(x+2)}{6}$ | $= \frac{3}{(x+2)} \times \frac{6}{(x+2)} = \frac{18}{(x+2)^2}$ |
| $\frac{222}{\left(\frac{1}{4}\right)}$ | $= \frac{222}{1} \times \frac{4}{1} = \frac{888}{1} = 888$ |

# Factoring

## Factoring Numbers and Expressions

### Prime Factorization

The **Prime Factorization** of a positive integer is the product of prime numbers. A **prime number** is an integer that is divisible by only by itself and 1. Some prime numbers are: $1, 2, 3, 5, 7, 11, 13, 17, ...$

Example: Find the prime factorization of 156

$156 = 2 \cdot 78$
$= 2 \cdot 2 \cdot 39$
$= 2 \cdot 2 \cdot 3 \cdot 13 \quad = \quad 2^2 \cdot 3 \cdot 13$

### Factoring Expressions

To factor an expression, identify the largest common factor and factor it out as shown below.

| | |
|---|---|
| $100x + 400$ | $= 100(x + 4)$ |
| $12x^2 + 4x + 10$ | $= 2(6x^2 + 2x + 5)$ |
| $8x^3 + 12x^2 + 16x$ | $= 4x(2x^2 + 3x + 4)$ |
| $a^2bc + ab^2cd + 2abc$ | $= abc(a + bd + 2)$ |

## Factoring Polynomials

Monomial   = Expression with one term
Binomial    = Expression with two terms
Trinomial   = Expression with three terms
Polynomial = Expression with one or more terms

When factoring polynomials, start by factoring out the greatest monomial. See previous page.

When factoring polynomials, the goal is to rearrange or rewrite the polynomial as a product of linear factors in the form: $a(x \pm b)(x \pm c)(x \pm d)$ ...

The notes and formulas in a later section of this book will provide guidance for factoring polynomials.

Meanwhile, here is an example.

$$3x^2 + 21x + 30 = 3(x^2 + 7x + 10)$$
$$= 3(x+2)(x+5)$$

| \multicolumn{2}{c}{**Factoring Formulas**} ||
|---|---|
| Perfect Square Trinomials | $(a+b)^2 = a^2 + 2ab + b^2$<br>$(a-b)^2 = a^2 - 2ab + b^2$ |
| Difference of 2 Squares | $a^2 - b^2 = (a-b)(a+b)$ |
| Sum and Diff. of 2 Cubes | $a^3 + b^3 = (a+b)(a^2 - ab + b^2)$<br>$a^3 - b^3 = (a-b)(a^2 + ab + b^2)$ |
| Distributive Property | $(a+b)(c+d+e)$<br>$= ac + ad + ae + bc + bd + be$ |
| Quadratic Formula | If: $\quad ax^2 + bx + c = 0$<br>Then: $\quad x = \dfrac{-b \pm \sqrt{b^2 - 4ac}}{2a}$ |
| Binomial Expansion | $(a+b)^n = \sum_{k=0}^{n} \binom{n}{k} a^{n-k} \cdot b^k$ |
| Combination | $\binom{n}{k} = \dfrac{n!}{k! \cdot (n-k)!}$ |
| Pascal's Triangle | 1<br>1   1<br>1   2   1<br>1   3   3   1<br>1   4   6   4   1<br>1   5   10   10   5   1 |

# Cartesian Coordinate System

## Graphing Numbers on a Number Line

One or more numbers can be represented on a number line. It may be necessary to estimate.

Example: Graph these numbers $\left\{-2, -\frac{1}{2}, \frac{5}{6}, \sqrt{7}, 5\right\}$

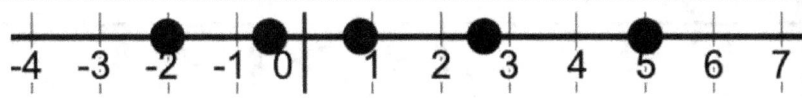

Hint: $\frac{5}{6}$ is almost $\frac{6}{6} = 1$

Hint: $\sqrt{7}$ is between $\sqrt{4} = 2$ and $\sqrt{9} = 3$

## Graphing Intervals on a Number Line

Intervals can be represented on a number line. Pay careful attention to the boundaries (ends).
- Open circle means the number is <u>not</u> included.
- Closed circle means the number is included.

| | |
|---|---|
| Numbers greater than −2 | |
| Numbers greater than 1 and less than or equal to 3. | |
| Numbers less than 2 and greater than 5 | |

## Graphing $(x, y)$ points on a Cartesian Plane

- The Cartesian Plane has two perpendicular axes.
- $x$ axis is the horizontal axis (Abscissa)
- $y$ axis is the vertical axis (Ordinate)
- $(x, y)$ is point on the cartesian plane.

| | |
|---|---|
| Graph the points: $(-2, 3), (2, 4), (3, -1)$ | 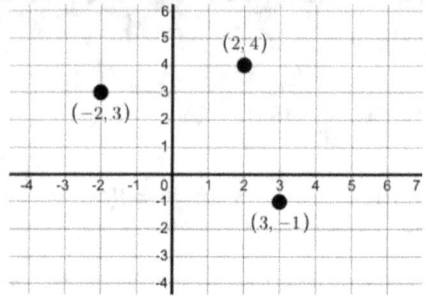 |
| Graph the points: $(-2, -2), (2, 2), (4, 4)$<br><br>Note: $y = x$ | 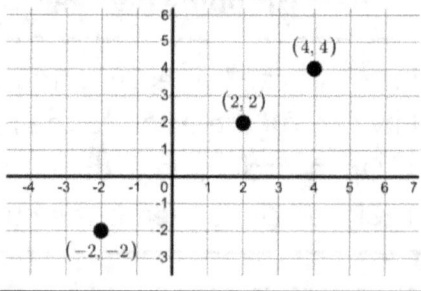 |
| Graph the points where $y = 2x$<br><br>Identify some points. $(-2, -4), (1, 2), (3, 6)$ | 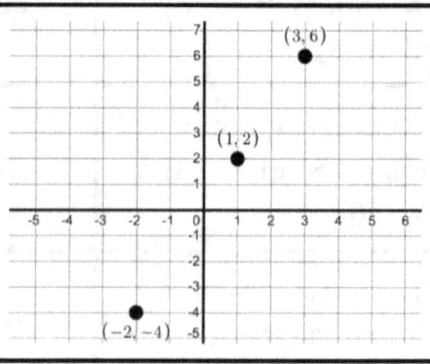 |

## Graphs on a Cartesian Plane

Graphs are like pictures of equations. How to make a graph for a given equation is shown below.
Given: $y = 2x + 1$

| | | | |
|---|---|---|---|
| Create a "T" table with some $x$ and $y$ values. | $x$ | $y = 2x + 1$ | $(x, y)$ |
| | 0 | 1 | $(0, 1)$ |
| | 1 | 3 | $(1, 3)$ |
| | 2 | 5 | $(2, 5)$ |
| | 3 | 7 | $(3, 7)$ |
| Plot the $(x, y)$ points | | | |
| Connect the dots. | | | |

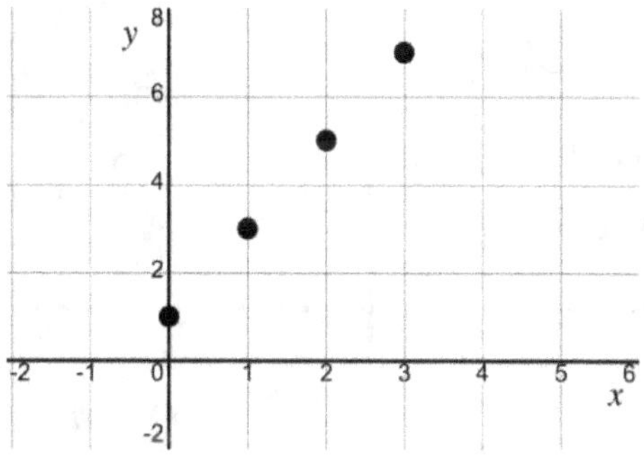

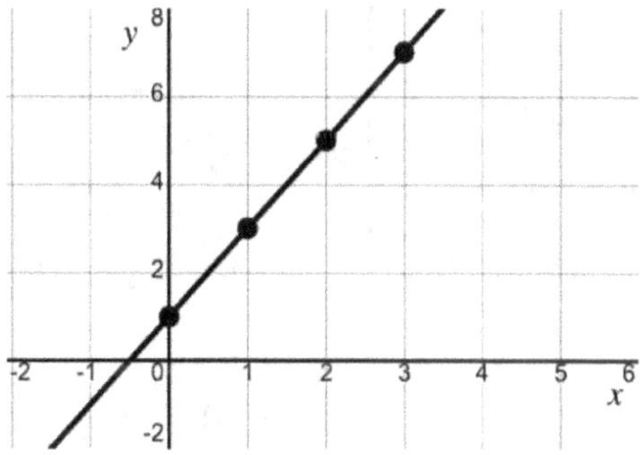

## Graphs – With a TI Calculator

How to make a graph for a given equation, on a TI graphing calculator, is shown below,
Given: $y = 2x + 1$

| | |
|---|---|
| Specify the x and y ranges. | Press [window] button. 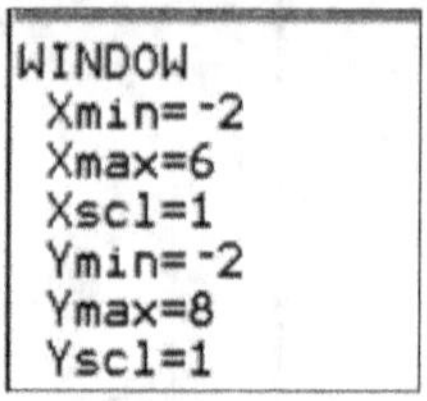<br><br>WINDOW<br>Xmin=-2<br>Xmax=6<br>Xscl=1<br>Ymin=-2<br>Ymax=8<br>Yscl=1 |
| Enter eqn.<br><br>Use [x,T,$\theta$,n] button for $x$ | Press [y=] button. 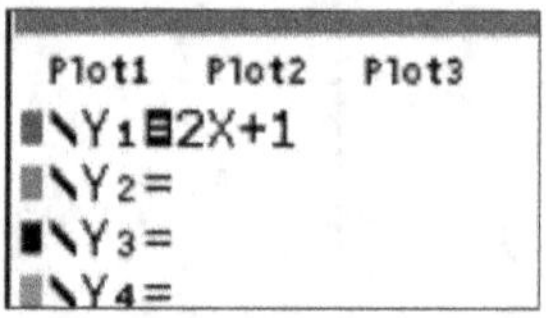<br><br>Plot1　Plot2　Plot3<br>\Y₁=2X+1<br>\Y₂=<br>\Y₃=<br>\Y₄= |
| Graph the equation. | Press [graph] button. 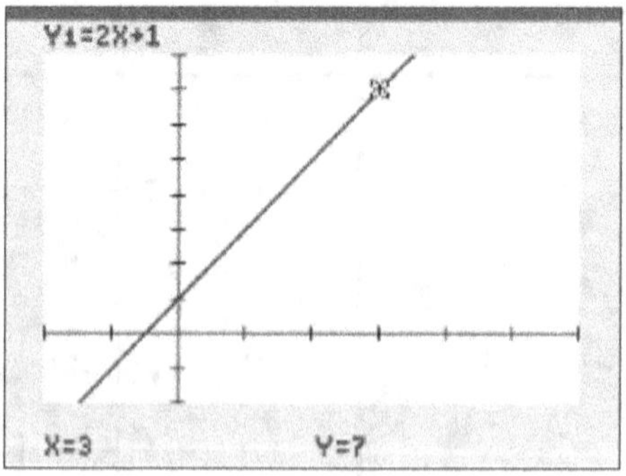<br><br>Y1=2X+1<br><br>X=3　　　Y=7<br>[trace], [enter] to see $x$ & $y$ values |

## Graphs – With Desmos

The Desmos app can be downloaded to your phone. Desmos is also available online (www.Desmos.com). Graphing an equation with Desmos is shown below.
Given: $y = 2x + 1$

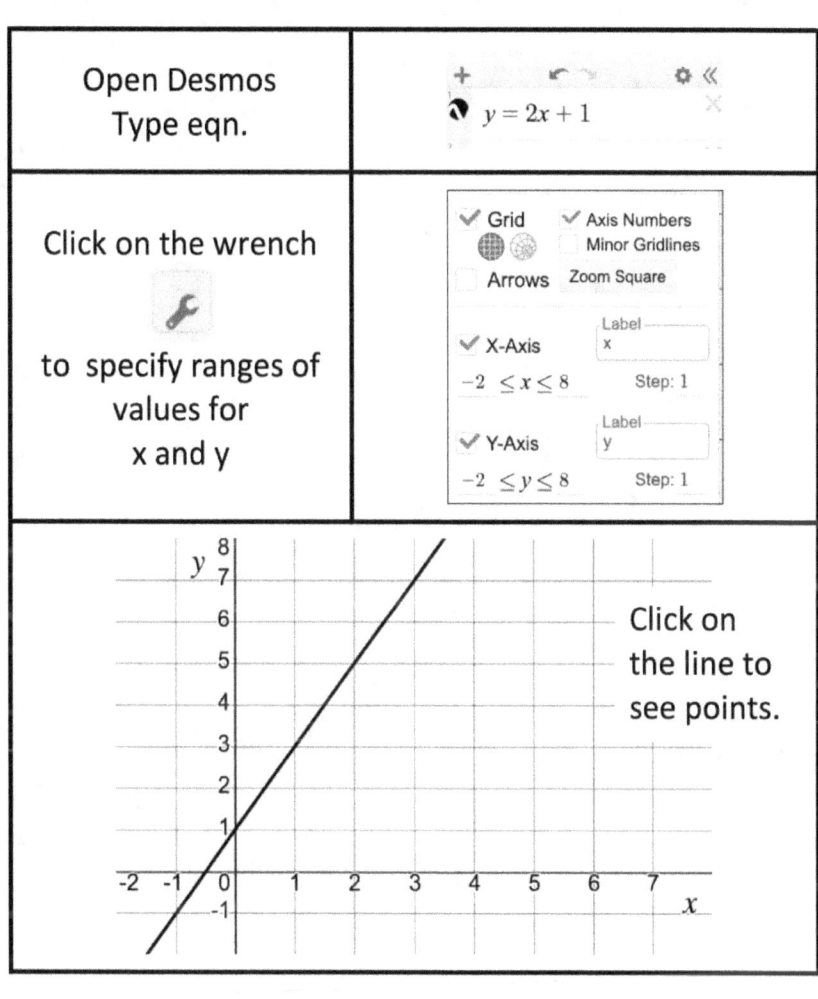

| Open Desmos Type eqn. | $y = 2x + 1$ |
| --- | --- |
| Click on the wrench to specify ranges of values for x and y | Grid, Axis Numbers, Minor Gridlines, Arrows, Zoom Square — X-Axis $-2 \leq x \leq 8$ Step: 1, Label x — Y-Axis $-2 \leq y \leq 8$ Step: 1, Label y |

Click on the line to see points.

# Expressions, Equations, Inequalities

## Expressions, Equations, and Inequalities
## Some Math Vocabulary

Some students confuse these terms. If you are one of those students, read this section. If you are not one of those students, feel free to skip this section.

This section includes:
- Expressions vs Equations
- Degree of an Expression or Equation
- Solving an Equation
- Equations vs Inequalities

## Math Expressions vs Equations

### Math Expressions

Expressions are just some math terms and operations. Expressions may be simplified but <u>not</u> solved.
Example:  $4x + 1$

### Math Equations

Equations are expressions with an equal sign. Equations may be simplified. In some cases equations may be rearranged to solve for a variable. When simplifying equations, always do the same thing to both sides of the equation. For example:

Solve for $x$:

$$4x + 1 = 9$$
$$4x + 1 - 1 = 9 - 1$$
$$4x = 8$$
$$\frac{4x}{4} = \frac{8}{4}$$
$$x = 2$$

## Degree of a Term, Expression or Equation

### Degree of a Term

- In general, the degree of a term is the exponent of the variable.
- If there is more than one variable, the degree is the sum of the exponents of the variables.
- If the term does not have a variable, the degree is 0
- The degree is always a non-negative integer.

| Some examples | $5x^3$ | Degree is 3 |
|---|---|---|
| | $2x^3y^5$ | Degree is $3 + 5 = 8$ |
| | $5x^{-3}$ | Not a term |
| | $5$ | Degree is 0<br>Note: $5 = 5x^0 = 5(1) = 5$ |

### Degree of an Expression or Equation

The degree of an expression or equation is the degree of the term with the highest degree. In standard form, terms are placed in order, based on their degree, highest to lowest. Consider the following expression...
$5x^3 + 7x^2 - 4x + 25$    Degree of expression is 3

## Solving Equations

Math equations with one or more variables are usually simplified (or rearranged) so that one variable is isolated. Examples below.

| | |
|---|---|
| $5x + 3 = 13$ <br> Solve for $x$ | $5x + 3 = 13$ <br> $5x = 10$ <br> $x = 2$ |
| $x^2 + 4 = 29$ <br> Solve for $x$ | $x^2 + 4 = 29$ <br> $x^2 = 25$ <br> $\sqrt{x^2} = \pm\sqrt{25}$ <br> $x = \pm 5$ |
| $y - x^2 + 5 = 8 + x$ <br> Solve for $y$ | $y - x^2 + 5 = 8 + x$ <br> $y + 5 = x^2 + x + 8$ <br> $y = x^2 + x + 3$ |
| $PV = nRT$ <br> Solve for $V$ | $PV = nRT$ <br> $V = \frac{nRT}{P}$ |
| $abc = 2b^2c$ <br> Solve for $a$ | $abc = 2b^2c$ <br> $a = \frac{2b^2c}{bc}$ <br> $a = 2b$ |

## Math Equations vs Inequalities

- **Equations** are expressions with an equal sign.
- **Inequalities** are expressions with an inequality sign.
- They may include zero, one, or more variables.
- They usually include one or two variables.

## Math Equation Examples

| $y = 2$ | $y = x + 1$ | $x = 3$ |
|---|---|---|
| 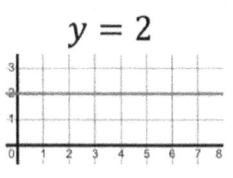 | 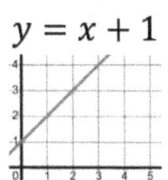 | 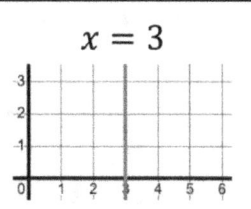 |

## Math Inequalities Examples

| Inequality Signs | > Greater than<br>< Less than<br>≤ Less than or equal<br>≥ Greater than or equal |
|---|---|

| $y < 2$ | $y \geq x + 1$ | $x \leq 2$ |
|---|---|---|
| 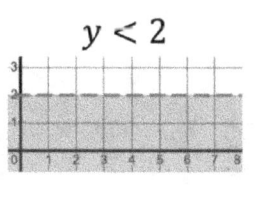 | 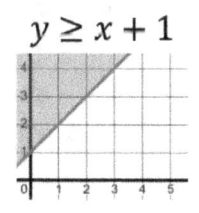 | 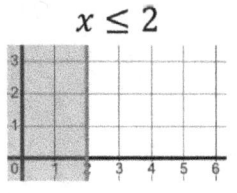 |

# Functions

## Functions

An equation with two variables (usually $x$ and $y$) represents a relationship between the two variables.

It can be rearranged to the form: $y = f(x)$
For example: 
$$2y + 3 = x + 4 + y$$
$$y + 3 = x + 4$$
$$y = x + 1$$
Now, $y = $ a function of $x$
$$y = f(x)$$

Here, $y$ is a function of one variable, $x$
The input to the function is $x$. The output is $y$.

If one input always results in the same output, the relationship is one-to-one $(1:1)$ and it is a **function**.

If one input results in one or more different outputs, the relationship is one-to-many and is NOT a function.

| Given $f(x)$ | Check to see if it is a function |
|---|---|
| $f(x) = x + 1$ | $f(5) = 6$     This is a function |
| $f(x) = \pm x + 1$ | $f(5) = 6, -4$     **Not** a function |

## How to Identify a Function from $(x, y)$ pairs

If given a set of $(x, y)$ pairs, we can check to see if they represent a function.

| $f(x) = x^2 + 1$ | Looking at the actual function and the list of $(x, y)$ pairs, we see that each $x$ input always give the same output. |
|---|---|
| x \| y <br> 0 \| 1 <br> 1 \| 2 <br> 2 \| 5 <br> 3 \| 10 <br> -2 \| 5 | Note that inputs of 2 and -2 give the same output. That's OK. An input of 2 always give the same output. An input of -2 always gives the same output. **This is a function.** |
| Function not given <br><br> x \| y <br> 0 \| 1 <br> 1 \| 2 <br> 2 \| 5 <br> 3 \| 10 <br> 1 \| 8 <br> -2 \| 4 | Looking at the list of $(x, y)$ pairs, we see that each $x$ input does NOT always give the same output. <br><br> An input of 1 gives an output of 2 or 8. <br> **This is NOT a function.** |

## How to Identify a Function Graphically

If given a graph of a function, we can easily check to see if each input ( $x$ ) gives just one output ( $y$ )

Use the **Vertical Line Test** to check.

$f(x) = x^2 + 1$

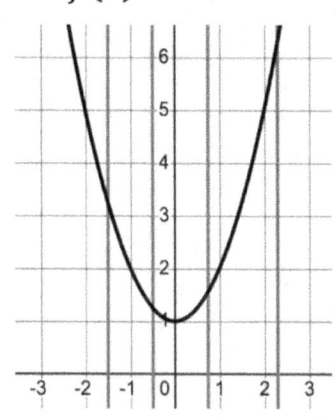

Vertical lines cross the graph in only one place.
This is a function.

Function not given.

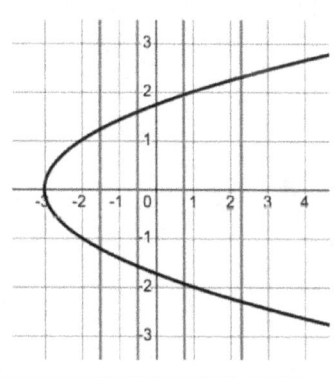

Vertical lines cross the graph in several places.
This is NOT a function.

# Linear and Quadratic Functions

## Degree of a Function

The degree of a function is the highest exponent of a variable in the function. For example:

$f(x) = 5x + 2$        Degree = 1
$f(x) = 5x + 2x^3 + 6$        Degree = 3
$f(x) = 7$        Degree = 0

## Linear Functions

Linear functions are 1ˢᵗ degree functions.
When graphed, they are straight lines (or linear).
The general form of a linear function is:

$$y = mx + b$$

$m =$ The slope of the line
$b =$ The y-intercept

## Quadratic Functions

Quadratic functions are 2ⁿᵈ degree functions.
When graphed, they are parabolas.
The general form of a quadratic function is:

$$y = ax^2 + bx + c$$

$a, b, c =$ constants
$a \neq 0$

| Linear Functions |
|---|
| Given two points: $(x_1, y_1)$ and $(x_2, y_2)$ One straight line can go through them. |

| Slope of straight line through 2 points. ||
|---|---|
| Slope $= m = \dfrac{\Delta y}{\Delta x} = \dfrac{Change\ in\ y}{Change\ in\ x} = \dfrac{y_2 - y_1}{x_2 - x_1}$ ||
| Parallel Lines | $m_2 = m_1$ |
| Perpendicular Lines | $m_2 = -\dfrac{1}{m_1}$ |

| Several Forms of a Linear Equation ||
|---|---|
| Slope-Intercept Form | $y = mx + b$ ; $b = y$ intercept |
| Point-Slope Form | $(\Delta y) = m(\Delta x)$ <br> $(y - y_1) = m(y - x_1)$ |
| Standard Form | $Ax + By = C$ |

## Quadratic Functions

### Several Forms of a Quadratic Equation

| Standard Form | $f(x) = ax^2 + bx + c$ |
|---|---|
| | Vertex: $\left(-\dfrac{b}{2a},\ f\left(-\dfrac{b}{2a}\right)\right)$ |
| Vertex Form | $f(x) = a(x - h)^2 + k$ |
| | Vertex: $(h, k)$ |
| Factored Form | $f(x) = a(x - z_1)(x - z_2)$ |
| | Vertex: $\left(\dfrac{z_1 + z_2}{2},\ f\left(\dfrac{z_1 + z_2}{2}\right)\right)$ |

### Special Product Formulas

$$(a + b)^2 = a^2 + 2ab + b^2$$

$$(a - b)^2 = a^2 - 2ab + b^2$$

$$(a + b)(a - b) = a^2 - b^2$$

Difference of Two Squares

## Linear Function Equations – Ex. 1a

Given two points: $(1, 5)$ and $(2, 8)$.
Find the equation of a straight line through them.

| | | |
|---|---|---|
| Find the slope | $m = \dfrac{\text{Change in } y}{\text{Change in } x} = \dfrac{\Delta y}{\Delta x} = \dfrac{y_2 - y_1}{x_2 - x_1}$ | |
| | $m = \dfrac{8-5}{2-1} = \dfrac{3}{1} = 3$ | |
| Point-Slope Form | $\Delta y = m \Delta x$ | |
| | $(y - 5) = 3(x - 1)$ | |
| Slope Intercept Form | $y = mx + b$ | |
| | $(y - 5) = 3(x - 1)$ | |
| | $y - 5 = 3x - 3$ | |
| | $y = 3x + 2$ | |
| Standard Form | $Ax + By = C$ | |
| | $y = 3x + 2$ | |
| | $-2 = 3x - y$ | |
| | $3x - y = -2$ | |

## Linear Function (Graph) – Ex. 1b

Given two points: $(1, 5)$ & $(2, 8)$ And the equation of the straight line through them. $y = 3x + 2$
Find the x-intercept and graph it.

| At the x-intercept, $y = 0$ | $y = 3x + 2$ $0 = 3x + 2$ $-2 = 3x$ $-\frac{2}{3} = x$ |
|---|---|
| | x-intercept at: $(x, y) = \left(-\frac{2}{3}, 0\right)$ |
| Graph | |

## Quadratic Function – Ex. 2

Given the quadratic equation: $y = 2(x-1)^2 - 5$
Convert it from vertex form to standard form.
Then, graph it.

| | |
|---|---|
| Expand | $y = 2(x-1)^2 - 5$ <br> $y = 2[(x-1)(x-1)] - 5$ <br> $y = 2[x^2 - x - x + 1] - 5$ <br> $y = 2[x^2 - 2x + 1] - 5$ <br> $y = 2x^2 - 4x + 2 - 5$ |
| Standard Form | $y = 2x^2 - 4x - 3$ |
| Graph | (1, -5) |

## Quadratic Function – Ex. 3

Given the quadratic equation: $y = x^2 - 4x + 3$

Convert it from standard form to vertex form:

$$y = a(x - h)^2 + k \quad ; \text{ Vertex at } (h, k)$$

| | | |
|---|---|---|
| Find $a$ | $a = 1 =$ coefficient of $x^2$ | |
| Find the vertex. | $x = -\dfrac{b}{2a} = \dfrac{4}{2} = 2$  $y = (2)^2 - 4(2) + 3 = -1$ | $h = 2$  $k = -1$ |
| Also, Graph Shows vertex | (graph of parabola with vertex at (2,-1)) | |
| Vertex Form | $y = (x - h)^2 + k$  $y = (x - 2)^2 - 1 \quad ;$ Vertex at $(2, -1)$ | |

## Midpoint & Distance – Ex. 4

Given two points: $(3, 2)$ and $(9, 10)$

Find the midpoint, on the straight line, between them.

Find the distance between the two points.

---

### Midpoint $= (Average\ x,\ Average\ y)$

Midpoint $= \left(\dfrac{x_1 + x_2}{2}, \dfrac{y_1 + y_2}{2}\right)$

Midpoint $= \left(\dfrac{3+9}{2}, \dfrac{2+10}{2}\right) = \left(\dfrac{12}{2}, \dfrac{12}{2}\right)$

Midpoint $= (6, 6)$

---

### Distance $= \sqrt{\Delta x^2 + \Delta y^2}$

Distance $= \sqrt{(x_2 - x_1)^2 + (y_2 - y_1)^2}$

Distance $= \sqrt{(9 - 3)^2 + (10 - 2)^2}$

Distance $= \sqrt{(6)^2 + (8)^2}$

Distance $= \sqrt{36 + 64}$

Distance $= \sqrt{100} = 10$

# Parent Functions

## Parent Functions

Parent functions are a set of different functions in simplest form. Students should know these functions and the graphs of them.

| $y = c$ | $y = c^x$ | $y = \log(x)$ |
|---|---|---|
| 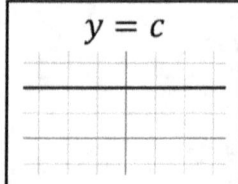 | 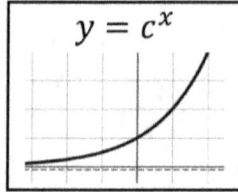 | 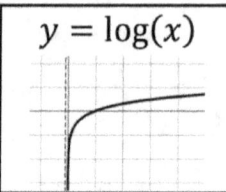 |
| $y = x$ | $y = |x|$ | $y = \dfrac{1}{x}$ |
| 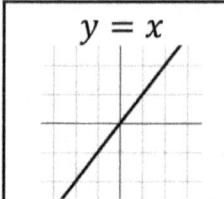 | 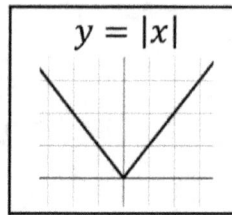 | 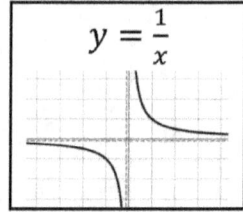 |
| $y = x^2$ | $y = \sqrt{x}$ | $y = \dfrac{1}{x^2}$ |
| 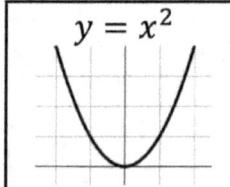 | 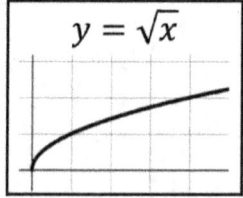 | 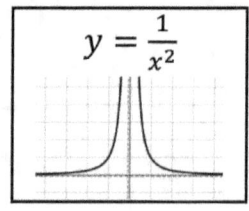 |
| $y = x^3$ | $y = \sqrt[3]{x}$ | $y = \lfloor x \rfloor$ |
| 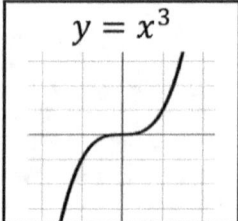 | 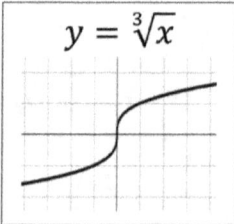 | 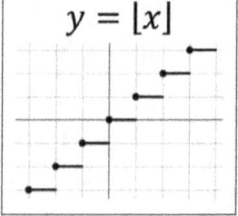 |

# Piecewise Functions

## Piecewise Function

A piecewise function has different functions for specified ranges of $x$. The general form of a piecewise function is:

$$y = \begin{cases} f_1(x) &, \quad x \leq a \\ f_2(x) &, \quad a < x \leq b \\ \ldots &, \quad \ldots \\ f_n(x) &, \quad x > n \end{cases}$$

### Piecewise Function Examples

| | |
|---|---|
| $y = \begin{cases} 1 &, \quad x \leq 0 \\ 3 &, \quad 0 < x \leq 2 \\ -1 &, \quad x > 2 \end{cases}$ | 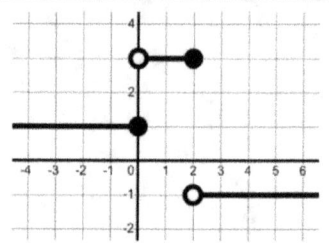 |
| $y = \begin{cases} 1 &, \quad x \leq 0 \\ x &, \quad x > 0 \end{cases}$ | 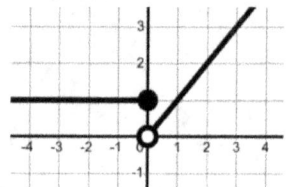 |
| $y = \begin{cases} 5 &, \quad x \leq -2 \\ x^2 &, \quad x > -2 \end{cases}$ | 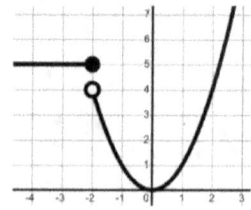 |

# Set Notation

> To specify a group of things, we use various types of set notation. In math, we usually use set notation to specify groups of numbers or ranges of numbers.

| Set Notation |  |
|---|---|
| A set of discrete numbers within parenthesis. ||
| { 2, 4, 6, 22, 123 } | Numbers: 2, 4, 6, 22, 123 |

| Set Builder Notation |  |
|---|---|
| A set of numbers within parenthesis. Provides more information than the simple set notation. ||
| { x \| 2 ≤ x < 33 } | The set, $x$, such that, $2 \leq x < 33$ |

| Interval Notation |  |
|---|---|
| Interval notation is a way to specify ranges of numbers, including the boundary conditions. A square bracket means the boundary is included. A round bracket means the boundary is not included. ||
| [ 1, 5 ] | Numbers between 1 and 5, including boundaries. |
| [ 1, 5 ) | Numbers between 1 and 5, including 1 but not 5. The numbers are less than 5. |

# Domain and Range

## Domain and Range of a Function $y = f(x)$

The **domain** is the set of valid inputs.
The **range** is the set of all possible outputs.

### Examples

| Function | Domain (x values) | Range (y values) |
|---|---|---|
| $y = x + 1$ | $(-\infty, \infty)$ Or we could say $x = \mathbb{R}$ | $(-\infty, \infty)$ Or we could say $y = \mathbb{R}$ |
| $y = x^2 - 3$ | $(-\infty, \infty)$ Or we could say $x = \mathbb{R}$ | $[-3, \infty)$ Or we could say $y \geq -3$ |
| $y = \dfrac{1}{x}$ | $(-\infty, 0) \cup (0, \infty)$ Or we could say $x \neq 0$ | $(-\infty, 0) \cup (0, \infty)$ Or we could say $y \neq 0$ |

## Domain and Range of a Function -- "Shade"

It may be helpful to
- Think of the domain as "shade" on the $x$ axis
- Think of the range as "shade" on the $y$ axis

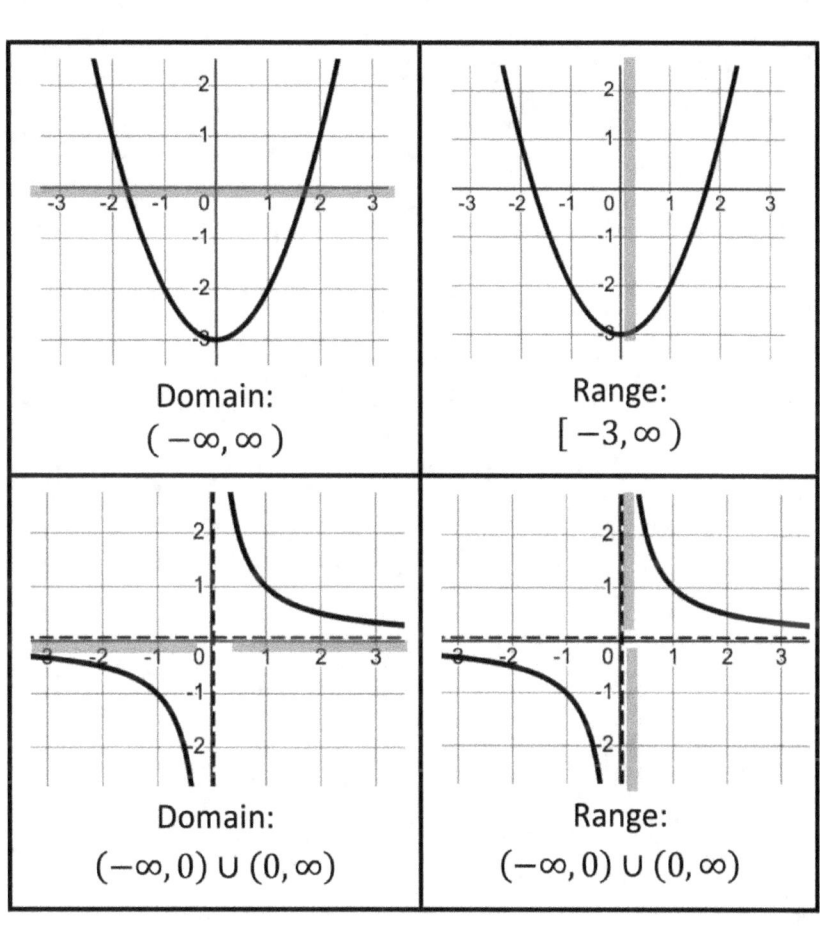

Domain:
$(-\infty, \infty)$

Range:
$[-3, \infty)$

Domain:
$(-\infty, 0) \cup (0, \infty)$

Range:
$(-\infty, 0) \cup (0, \infty)$

# More Linear Equations

| **Linear Function Review** |
|---|
| Given two points: $(x_1, y_1)$ and $(x_2, y_2)$ |
| One straight line can go through them. |

| Slope of straight line through 2 points. ||
|---|---|
| Slope $= m = \dfrac{\Delta y}{\Delta x} = \dfrac{Change\ in\ y}{Change\ in\ x} = \dfrac{y_2 - y_1}{x_2 - x_1}$ ||
| Parallel Lines | $m_2 = m_1$ |
| Perpendicular Lines | $m_2 = -\dfrac{1}{m_1}$ |

| Several Forms of a Linear Equation ||
|---|---|
| Slope-Intercept Form | $y = mx + b \quad ; \ b = y$ intercept |
| Point-Slope Form | $(\Delta y) = m(\Delta x)$<br>$(y - y_1) = m(y - x_1)$ |
| Standard Form | $Ax + By = C$ |

## Linear Function – Ex. 1

Given 2 Points: $(2, 5)$ and $(4, -3)$
Find equation of a line that passes through them.

| | |
|---|---|
| Find Slope | $m = \dfrac{\Delta y}{\Delta x}$ <br><br> $m = \dfrac{-3 - 5}{4 - 2} = \dfrac{-8}{2}$ <br><br> $m = -4$ |
| Point-Slope Form | $\Delta y = m \Delta x$ <br><br> $(y - 5) = -4(x - 2)$ |
| Slope-Intercept Form | $y = mx + b$ <br><br> $(y - 5) = -4(x - 2)$ <br><br> $y - 5 = -4x + 8$ <br><br> $y = -4x + 13$ |
| Standard Form | $Ax + By = C$ <br><br> $4x + y = 13$ |

## Linear Function – Ex. 2

Given 1 Point: $(1, 3)$ and a slope $= 5$

Find equation of a straight line with the given slope and point.

| Find Slope | Given. $m = 5$ |
|---|---|
| Point-Slope Form | $\Delta y = m\Delta x$ |
| | $(y - 3) = 5(x - 1)$ |
| Slope-Intercept Form | $y = mx + b$ |
| | $(y - 3) = 5(x - 1)$ <br> $y - 3 = 5x - 5$ <br> $y = 5x - 2$ |
| Standard Form | $Ax + By = C$ |
| | $y = 5x - 2$ <br> $y + 2 = 5x$ <br> $2 = 5x - y$ <br> $5x - y = 2$ |

## Linear Function – Ex. 3

Given an equation: $y = 4x - 5$

Find equation of a straight line that is perpendicular to the given line and passes through point $(2, 3)$

| | | |
|---|---|---|
| Find Slope | $m = -\frac{1}{4}$ | |
| Point-Slope Form | $\Delta y = m\Delta x$ | |
| | $(y - 3) = -\frac{1}{4}(x - 2)$ | |
| Slope-Intercept Form | $y = mx + b$ | |
| | $(y - 3) = -\frac{1}{4}(x - 2)$ | |
| | $y - 3 = -\frac{1}{4}x + \frac{1}{2}$ | |
| | $y = -\frac{1}{4}x + \frac{7}{2}$ | |
| Standard Form | $Ax + By = C$ | |
| | $y = -\frac{1}{4}x + \frac{7}{2}$ | |
| | $\frac{1}{4}x + y = \frac{7}{2}$ | |
| | $x + 4y = 14$ | |

## Linear Function – Ex. 4

Find equation of a straight line that crosses the x-axis at $x = -5$ and crosses the y-axis at $y = 2$

| | |
|---|---|
| Identify the two points | Points: $(-5, 0)$ and $(0, 2)$ |
| Find Slope | $m = \dfrac{\Delta y}{\Delta x} = \dfrac{2-0}{0-(-5)} = \dfrac{2}{5}$ |
| Point-Slope Form | $\Delta y = m \Delta x$ |
| | $(y - 2) = \dfrac{2}{5}(x - 0)$ |
| Slope-Intercept Form | $y = mx + b$ |
| | $(y - 2) = \dfrac{2}{5}(x)$ |
| | $y = \dfrac{2}{5}x + 2$ |
| Standard Form | $Ax + By = C$ |
| | $y = \dfrac{2}{5}x + 2$ |
| | $-\dfrac{2}{5}x + y = 2$ |
| | $-2x + 5y = 10$ |

## Linear Function – Ex. 5

Given the point $(4, -6)$

Find:
- A horizontal line that passes through the point.
- A vertical line that passes through the point.

| Vertical Line | $x = 4$ |
|---|---|
| Horizontal Line | $y = -6$ |

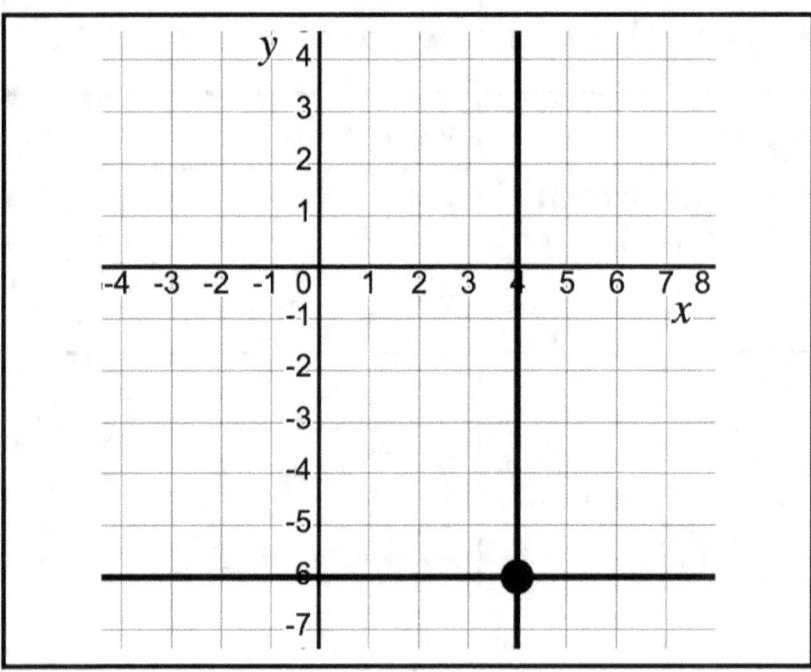

| **Linear Function – Ex. 6** |
|---|
| Given: $y = 2x + 3$ |
| Find: The $x$ and $y$ intercepts. |

| | |
|---|---|
| | At $x$ intercept, $y = 0$ |
| $x$ intercept | $(0) = 2x + 3$ <br> $-3 = 2x$ <br> $-\dfrac{3}{2} = x$ |
| | Point: $\left(-\dfrac{3}{2}, 0\right)$ |

| | |
|---|---|
| | At $y$ intercept, $x = 0$ |
| $y$ intercept | $y = 2(0) + 3$ <br> $y = 0 + 3$ <br> $y = 3$ |
| | Point: $(0, 3)$ |

## Linear Function – Ex. 7

Find the intersection point of two lines.
Given: $y_1 = 2x + 5$ and $y_2 = -3x + 10$

| | |
|---|---|
| Set equations equal | $y_1 = y_2$ <br> $2x + 5 = -3x + 10$ |
| Solve for $x$ | $5x + 5 = 10$ <br> $5x = 5$ <br> $x = 1$ |
| Find the $y$ That goes with $x = 1$ | $y = 2x + 5$ <br> $y = 2(1) + 5$ <br> $y = 7$ <br><br> Point: $(x, y) = (1, 7)$ |
| Graph | |

## Midpoint & Distance Between Two Points – Ex. 8

Given 2 Points: $(2, 5)$ and $(4, -3)$

Find the midpoint and distance between them.

| Midpoint | $(Avg.\,x, Avg.\,y) = \left(\dfrac{x_1 + x_2}{2}, \dfrac{y_1 + y_2}{2}\right)$ |
|---|---|
| | $\left(\dfrac{2+4}{2}, \dfrac{5+(-3)}{2}\right) = \left(\dfrac{6}{2}, \dfrac{2}{2}\right) = (3, 1)$ |

| Distance between the points | $d = \sqrt{(\Delta x)^2 + (\Delta y)^2}$ |
|---|---|
| | $d = \sqrt{(4-2)^2 + (-3-5)^2}$ |
| | $d = \sqrt{(2)^2 + (-8)^2}$ |
| | $d = \sqrt{4 + 64}$ |
| | $d = \sqrt{68}$ |
| | $d = \sqrt{4 \cdot 17} = \sqrt{4} \cdot \sqrt{17}$ |
| | $d = 2\sqrt{17}$ |

# Absolute Value Problems

| Absolute Value Problems |
|---|
| $$\|x\| = \begin{cases} x & , x \geq 0 \\ -x & , x < 0 \end{cases}$$ |
| The output from an absolute value function is always positive. |
| $\|x\| = $ a negative number is **UNDEFINED**. |

## Absolute Value Equations

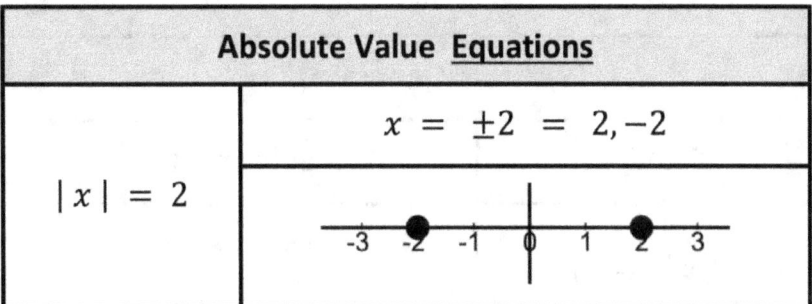

| $\|x\| = 2$ | $x = \pm 2 = 2, -2$ |
|---|---|

## Absolute Value Inequalities

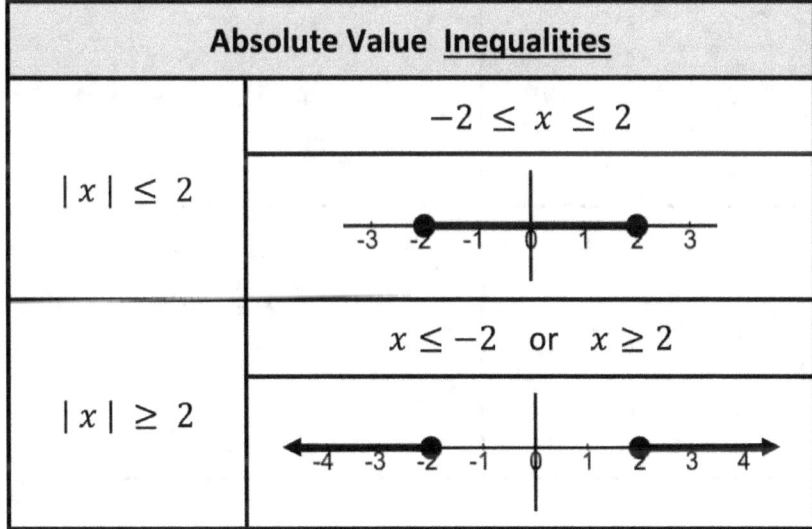

| $\|x\| \leq 2$ | $-2 \leq x \leq 2$ |
|---|---|
| $\|x\| \geq 2$ | $x \leq -2$ or $x \geq 2$ |

| Absolute Value Equation – Simple Examples ||
|---|---|
| Problem | Solve for x |
| $\|x\| = 13$ | $\|x\| = 13$ <br> $x = \pm 13$ <br> $x = 13, -13$ |

| In General ||
|---|---|
| Problem | Solve for x |
| $\|x\| = n$ | $\|x\| = n$ <br> $x = \pm n$ |
| $\|x + a\| = n$ | $\|x + a\| = n$ <br> $x + a = \pm n$ <br> $x = -a \pm n$ |
| $\|x + 2\| = 9$ | $\|x + 2\| = 9$ <br> $x + 2 = \pm 9$ <br> $x = -2 \pm 9$ <br> $x = 7, -11$ |

| Absolute Value Inequality – Simple Examples ||
|:---:|:---:|
| Problem | Solution Format |
| $\|x\| \leq 3$ | closed interval $[-3, 3]$ |
| $\|x\| \geq 3$ | $(-\infty, -3] \cup [3, \infty)$ |

| In General ||
|:---:|:---:|
| Problem | Graph |
| $\|x\| \leq n$ | closed interval $[-n, n]$ |
| $\|x\| \geq n$ | $(-\infty, -n] \cup [n, \infty)$ |

| Absolute Value Equation – Ex. 0 |
|---|
| Given: $\|x + 3\| = -13$ <br> Solve for $x$ |

| Solution | **No Solution** <br><br> The output from an absolute value function is ALWAYS POSITIVE. <br> It cannot be negative. |
|---|---|
| Check a few numbers. <br><br> Try $x = 16$ | $\|x + 3\| = -13$ <br> $\|16 + 3\| = -13$ <br> $\|19\| = -13$ <br> $19 = -13$    **FALSE** |
| Try <br> $x = -16$ | $\|x + 3\| = -13$ <br> $\|-16 + 3\| = -13$ <br> $\|-13\| = -13$ <br> $13 = -13$    **FALSE** |

| Absolute Value Equation – Ex. 1 |
|---|
| Given: $\|x+3\| = 13$<br>Solve for $x$ |

| | |
|---|---|
| Solution | $\|x+3\| = 13$<br>$x + 3 = \pm 13$<br>$x = -3 \pm 13$<br>$x = 10, -16$ |
| Check<br>$x = 10$ | $\|x+3\| = 13$<br>$\|10 + 3\| = 13$<br>$\|13\| = 13$<br>$13 = 13$    TRUE |
| Check<br>$x = -16$ | $\|x+3\| = 13$<br>$\|-16 + 3\| = 13$<br>$\|-13\| = 13$<br>$13 = 13$    TRUE |

| Absolute Value Equation – Ex. 2 |
|---|
| Given: $\mid 2x + 3 \mid = 13$ <br> Solve for $x$ |

| | |
|---|---|
| Solution | $\mid 2x + 3 \mid = 13$ <br> $2x + 3 = \pm 13$ <br> $2x = -3 \pm 13$ <br> $2x = 10, -16$ <br> $x = 5, -8$ |
| Check <br> $x = 5$ | $\mid 2x + 3 \mid = 13$ <br> $\mid 10 + 3 \mid = 13$ <br> $\mid 13 \mid = 13$ <br> $13 = 13$   TRUE |
| Check <br> $x = -8$ | $\mid 2x + 3 \mid = 13$ <br> $\mid -16 + 3 \mid = 13$ <br> $\mid -13 \mid = 13$ <br> $13 = 13$   TRUE |

## Absolute Value Inequality – Ex. 3

Given: $|2x + 3| \leq 13$

Solve for $x$. Give your answer in interval notation.

| | |
|---|---|
| One range of values | $-13 \leq (2x + 3) \leq 13$ <br> $-16 \leq 2x \leq 10$ <br> $-8 \leq x \leq 5$ |
| Solution | $x = [-8, 5]$ |

## Absolute Value Inequality – Ex. 4

Given: $|2x + 3| > 13$

Solve for $x$. Give your answer in interval notation.

| | |
|---|---|
| Two ranges of values | $(2x + 3) > 13$ <br> $2x > 10$ <br> $x > 5$ |
| | $(2x + 3) < -13$ <br> $2x < -16$ <br> $x < -8$ |
| Solution | $x = (-\infty, -8) \cup (5, \infty)$ |

| Absolute Value Inequality – Ex. 5a |
|---|
| Given: $|5-x| > 13$ <br> Solve for $x$. Give your answer in interval notation. |

| | |
|---|---|
| Two ranges of values <br><br> **NOTE:** <br> When multiplying an inequality by a negative number, **FLIP** the inequality sign | $(5-x) > 13$ <br> $-x > 8$ <br> $x < -8$ |
| | $(5-x) < -13$ <br> $-x < -18$ <br> $x > 18$ |
| Solution | $x = (-\infty, -8) \cup (18, \infty)$ <br><br> ⬅──○────○──➡ <br>    -8     18 |

## Absolute Value Inequality – Ex. 5b
### ( Different Solution, Same Answer! )

Given: $|5-x| > 13$

Solve for $x$. Give your answer in interval notation.

---

With this solution, we avoid multiplying the inequality by a negative sign. So, there is no need to flip the inequality sign. Start by adding $x$ to both sides.

| | |
|---|---|
| Two ranges of values | $(5-x) > 13$ <br> $5 > 13 + x$ <br> $-8 > x$ |
| | $(5-x) < -13$ <br> $5 < -13 + x$ <br> $18 < x$ |
| Solution | $x = (-\infty, -8) \cup (18, \infty)$ <br><br> ⟵──o────o──⟶ <br>     -8     18 |

# Scientific Notation

**Scientific Notation Format:** $m \times 10^n$
$n$ = integer, $m$ = real

Examples:
- $1230000 = 123 \times 10^4$
- $1230000 = 12.3 \times 10^5$
- $1230000 = 1.23 \times 10^6$ (Standard Form)
- $1230000 = 0.123 \times 10^7$

Standard form is when: $1 \leq m < 10$
There is just one digit to the left of the decimal point.

| Scientific Notation for Large Numbers (examples) ||
|---|---|
| 1 million | $= 1{,}000{,}000 = 1.0 \times 10^6$ |
| 1,206.5 | $= 1.2065 \times 10^3$ |

| Scientific Notation for Small Numbers (examples) ||
|---|---|
| 0.005 | $= 5.0 \times 10^{-3}$ |
| $\dfrac{456}{10000}$ | $= .0456 = 4.56 \times 10^{-2}$ |

## Scientific Notation – Ex. 1

Write the given numbers in scientific notation.

Note: On a TI calculator, the [mode] can be used to set the exponential format to Scientific Notation. Scientific Notation, on a TI calculator is: $mEn$

| Number | Scientific Notation |
|---|---|
| 5799 | $= 5.799 \times 10^3 \quad = \quad 5.799E3$ |
| 40,090 | $= 4.009 \times 10^4$ |
| 7,622,522 | $= 7.622522 \times 10^6$ |
| 2,888,000,000 | $= 2.888 \times 10^9$ |
| 0.004 | $= 4.0 \times 10^{-3}$ |
| 0.0502 | $= 5.02 \times 10^{-2}$ |
| 42 | $= 4.2 \times 10^1$ |
| $\frac{5}{4}$ | $= 1.25$ |
| $\frac{4}{5}$ | $= 0.8 = 8.0 \times 10^{-1}$ |

## Scientific Notation – Ex. 2

Write the given numbers in decimal form.

| Number | Decimal Form |
|---|---|
| $3.45 \times 10^3$ | $= 3450.0$ |
| $3.45 \times 10^5$ | $= 345000.0$ |
| $3.45 \times 10^{-3}$ | $= 0.00345$ |
| $3.45 \times 10^{-5}$ | $= 0.0000345$ |
| $7.0 \times 10^9$ | $= 7,000,000,000.0$ |
| $4.5E2$ | $= 450.0$ |
| $1.234E5$ | $= 123400.0$ |
| $1.234E(-5)$ | $= 0.00001234$ |
| $1.11 \times 10^2$ | $= 111.0$ |
| $1.11 \times 10^{-2}$ | $= 0.0111$ |
| $1.1212 \times 10^{-6}$ | $= 0.0000011212$ |
| $1.1212 \times 10^6$ | $= 1121200.0$ |

## Scientific Notation – Ex. 3

Write the given numbers in scientific notation.

| Number | Scientific Notation |
|---|---|
| The radius of the earth's orbit is $150{,}000{,}000{,}000\ m$ | $1.5 \times 10^{11}$ |
| The diameter of a red blood cell is about $0.00074\ cm$ | $7.4 \times 10^{-4}$ |
| A communications satellite was orbited at an altitude of $725{,}000\ m$ | $7.25 \times 10^5$ |
| The diameter of the earth is 7,926 miles. Express this number in feet and in scientific notation. Round to 4 significant digits. Use: $1\ mile = 5280\ ft.$ | $7926\ mi. \left(\frac{5280\ ft.}{1\ mi.}\right)$ $= 41849280\ ft.$ $= 4.185 \times 10^7\ ft.$ |

# Making Choices – Bike Rental

## Making Choices
## Bike Rental Example

Math equations can be very useful. For example, math equations can help decide between two options.

In this section, math equations will be used to model and compare two options.

The two options will be compared in multiple ways:
- Algebraically
- Graphically
- Using Technology
    - TI Graphing Calculator
    - Desmos Graphing app

| **Bike Rental – Ex. 1a** |
|---|
| Suppose you wanted to rent a bike for one day, from one of two bike rental companies.<br><br>Bike Rental Company A:   $8/day plus  10 cents/mile<br>Bike Rental Company B:   $4/day plus  50 cents/mile |
| Write equations to represent the total cost of riding a bike for x miles, for each bike company.<br><br>Hint: Convert the cents to dollars so the money units match (all in dollars).  For example:  5 cents = $0.05 |

| |
|---|
| Cost A  =  $8  +  $0.10 · $x$ |
| Cost B  =  $4  +  $0.50 · $x$ |

## Bike Rental – Ex. 1b

Use the equations, previously found, to calculate the cost if for riding a rental bike, from both bike companies, for 5, 10, 15, 20 miles.

| Previously found | Cost A = $8 + $0.10 · x |
|---|---|
| | Cost B = $4 + $0.50 · x |

| $x$ | Company | Cost | |
|---|---|---|---|
| 5 | A | 8 + .10(5) | = $ 8.50 |
| | B | 4 + .50(5) | = $ 6.50 |
| 10 | A | 8 + .10(10) | = $ 9.00 |
| | B | 4 + .50(10) | = $ 9.00 |
| 15 | A | 8 + .10(15) | = $ 9.50 |
| | B | 4 + .50(15) | = $ 11.50 |
| 20 | A | 8 + .10(20) | = $ 10.00 |
| | B | 4 + .50(20) | = $ 14.00 |

## Bike Rental – Ex. 1c

Use the equations, previously found, to calculate the break-even point. In other words, find the number of miles $(x)$ that makes both equations equal.

| Previously found | Cost A = \$8 + \$0.10 · $x$ <br> Cost B = \$4 + \$0.50 · $x$ |
|---|---|
| Set equations equal. | $8 + 0.10x = 4 + 0.50x$ |
| Solve for $x$ | $8 + 0.10x = 4 + 0.50x$ <br> $4 + 0.10x = 0.50x$ <br> $4 = 0.40x$ <br> $\dfrac{4}{.40} = x$ <br> $10 = x$ |
| Break Even Point | When $x = 10$ both equations are equal. |
| Conclusion | If you plan to ride more than 10 miles, rent from company A. |

## Bike Rental – Ex. 1d

Graph the equations, previously found, to calculate the break-even point. That's where the two graphs intersect, and are equal.

| Previously found | Cost A = $8 + $0.10 · x |
| | Cost B = $4 + $0.50 · x |

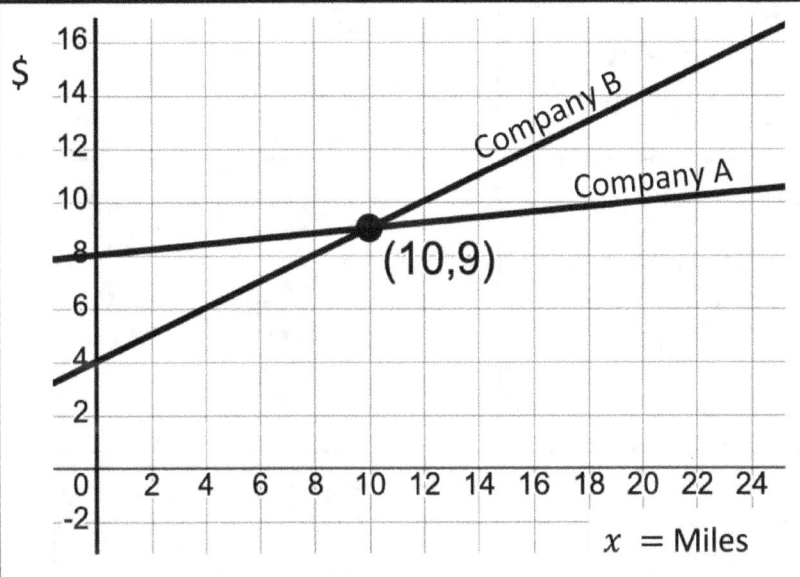

The break-even point is $(x, y) = (10, 9)$ which means that both companies will charge $9 for 10 miles.
A is a better choice if riding more than 10 miles.

| Bike Rental – Calculator Graph |
|---|
| Graph the equations, previously found, to calculate the break-even point. Use a TI graphing calculator. |

| Previously found | Cost A = $8 + $0.10 \cdot x$ <br> Cost B = $4 + $0.50 \cdot x$ |
|---|---|

| Enter both equations into the calculator. <br><br> Use [x,T,θ,n] for $x$ | Press [y=] button. <br><br> Plot1  Plot2  Plot3 <br> ■\Y₁=8+.10X <br> ■\Y₂=4+.50X <br> ■\Y₃= |
|---|---|
| Use window to specify x and y ranges. | Press [window] button. <br><br> WINDOW <br> Xmin=-2 <br> Xmax=22 <br> Xscl=1 <br> Ymin=-2 <br> Ymax=20 <br> Yscl=1 |
| Graph equations. | Press [graph] button. [trace] <br><br> 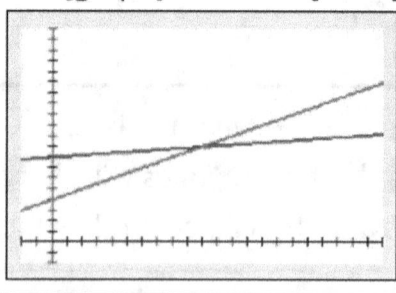 |

## Bike Rental – Calculator Graph Continued

Graph the equations, previously found, to calculate the break-even point. Use a TI graphing calculator.

| Previously found | Cost A $= \$8 + \$0.10 \cdot x$ <br> Cost B $= \$4 + \$0.50 \cdot x$ |
|---|---|

| Find the intersection. | Press [2$^{nd}$], [trace] to calculate. Select "intersect", [enter] |
|---|---|
| Answer questions. | 1$^{st}$ curve? [enter] <br> 2$^{nd}$ curve? [enter] <br> Guess? [enter] |
| Intersection at $(x, y) = (10, 9)$ | Y2=4+.50X <br><br> Intersection <br> X=10    Y=9 |

The break-even point is $(x, y) = (10, 9)$ which means that both companies will charge $9 for 10 miles.

| **Bike Rental – Desmos Graph** |
|---|
| Graph the equations, previously found, to calculate the break-even point. Use Desmos.<br><br>The Desmos app can be downloaded to your phone. Or, Desmos can be used online (www.Desmos.com) |

| Previously found | Cost A = \$8 + \$0.10 · x<br>Cost B = \$4 + \$0.50 · x |
|---|---|
| Type the two equations. | $y = 8 + .10x$<br>$y = 4 + .50x$ |
| Click the wrench.  to setup ranges for x and y | 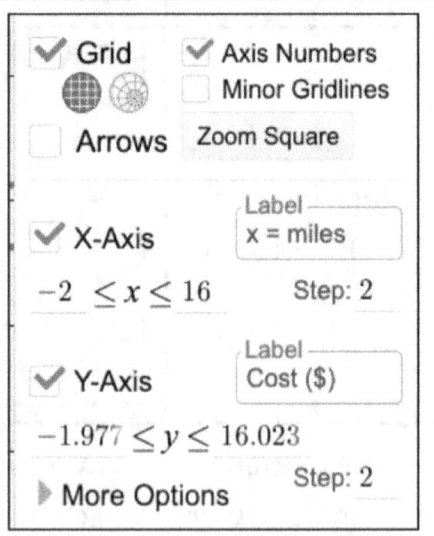 |

## Bike Rental – Desmos Graph Continued

Graph the equations, previously found, to calculate the break-even point. Use Desmos.

The Desmos app can be downloaded to your phone. Or, Desmos can be used online (www.Desmos.com)

| Previously found | Cost A = \$8 + \$0.10 · x<br>Cost B = \$4 + \$0.50 · x |
|---|---|

Graph is displayed.

Tap on the intersection to see the point.

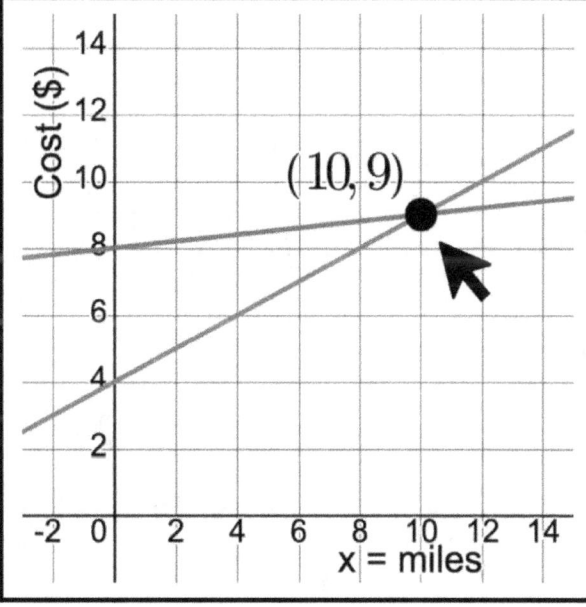

# **Algebra Topics - Part 1**

# Polynomial Division (Long & Synthetic)

## Division -- Review

$$\frac{Dividend}{Divisor} = Quotient \quad \rightarrow \quad Divisor\overline{\smash{)}Dividend}^{\,Quotient}$$

### Regular Division with Numbers (Quick Review)

$\dfrac{425}{5}$

$= 85$

```
        8  5
   5  ) 4  2  5
        4  0
        ─────
           2  5
           2  5
           ─────
              0   Remainder
```

$\dfrac{426}{5}$

$= 85 + \dfrac{1}{5}$

```
        8  5
   5  ) 4  2  6
        4  0
        ─────
           2  6
           2  5
           ─────
              1   Remainder
```

## Division of Polynomials

- **Long division** of polynomials is similar to regular division. No restrictions on the divisor.
- **Synthetic division** is only used with 1<sup>st</sup> degree divisors in the form $(x - c)$.

| | |
|---|---|
| For comparison, the same problem is done with long and synthetic division, below. | $$\frac{x^3 + 3x^2 - x - 3}{x - 1}$$ |
| Long Division | $$\begin{array}{r} x^2 + 4x + 3 \\ (x-1) \overline{\smash{)} x^3 + 3x^2 - x - 3} \\ \underline{-(x^3 - x^2)} \phantom{xxxxxxxxx} \\ 4x^2 - x \phantom{xxx} \\ \underline{-(4x^2 - 4x)} \phantom{xx} \\ 3x - 3 \\ \underline{-(3x - 3)} \\ \text{Remainder} \quad 0 \end{array}$$ |
| Synthetic Division | $$\begin{array}{c|cccc} \boxed{1} & 1 & 3 & -1 & -3 \\ & & 1 & 4 & 3 \\ \hline & 1 & 4 & 3 & 0 \end{array}$$ |
| Answer | $x^2 + 4x + 3 + \dfrac{0}{x-1}$ |

## Division of Polynomials – Ex. 01
### Long Division

Use long division to evaluate: $\dfrac{x^3}{x^2+1}$

Note: All terms must be represented. (add fillers)

$$(x^2 + 0x + 1) \overline{\smash{\big)}\ \begin{array}{l} x \\ x^3 + 0x^2 + 0x + 0 \\ -\underline{(x^3 + 0x^2 + 1x\,)} \\ \phantom{x^3 + 0x^2 + {}} -x \end{array}}$$

Remainder (← $-x$)

**Answer:** $\dfrac{x^3}{x^2+1} = x - \dfrac{x}{x^2+1}$

## Division of Polynomials – Ex. 02
## Long Division & Synthetic Division

Use <u>long division</u> and <u>synthetic division</u> to evaluate:

$$\frac{x^2 + x^3 - 2x - 5}{x - 3}$$

Note: Put terms in proper order.

Solution

$$
(x - 3) \begin{array}{r} x^2 + 4x + 10 \\ \overline{\smash{)}x^3 + x^2 - 2x - 5} \\ -\underline{(x^3 - 3x^2)} \phantom{XXXXXXX} \\ 4x^2 - 2x \phantom{XXX} \\ -\underline{(4x^2 - 12x)} \phantom{XX} \\ 10x - 5 \\ -\underline{(10x - 30)} \\ 25 \end{array}
$$

Remainder

Put the zero in the box.

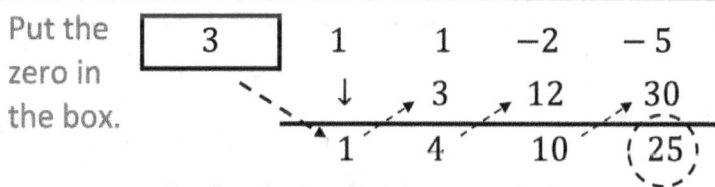

Answer $\quad \dfrac{x^2 + x^3 - 2x - 5}{x - 3} = x^2 + 4x + 10 + \dfrac{25}{x - 3}$

## Division of Polynomials – Ex. 03
## Long Division & Synthetic Division

| Use <u>long</u> and <u>synthetic</u> <u>division</u> to evaluate. | $\dfrac{2x^4 - x^3 + 4x^2 - 14x + 6}{2x - 1}$ |

**Solution**

$$
\begin{array}{r}
x^3 + 2x - 6 \phantom{xxxxxx} \\
(2x-1) \overline{\smash{\big)}\, 2x^4 - x^3 + 4x^2 - 14x + 6} \\
\underline{-\;(2x^4 - x^3)\phantom{xxxxxxxxxxxx}} \\
0 + 4x^2 - 14x \phantom{xx} \\
\underline{-\;(4x^2 - 2x)\phantom{xxx}} \\
-12x + 6 \\
\underline{-\;(-12x + 6)} \\
0
\end{array}
$$

$$\dfrac{2x^4 - x^3 + 4x^2 - 14x + 6}{2x - 1} = \dfrac{x^4 - \tfrac{1}{2}x^3 + 2x^2 - 7x + 3}{x - \tfrac{1}{2}}$$

| $\boxed{\tfrac{1}{2}}$ | 1 | $-\tfrac{1}{2}$ | 2 | $-7$ | 3 |
|---|---|---|---|---|---|
|  | ↓ | $\tfrac{1}{2}$ | 0 | 1 | $-3$ |
|  | 1 | 0 | 2 | $-6$ | 0 |

$$\dfrac{2x^4 - x^3 + 4x^2 - 14x + 6}{2x - 1} = x^3 + 2x - 6$$

# Factoring Polynomials

**Factoring Binomials:** $a^2 - b^2 = (a-b)(a+b)$
**The Difference of Two Squares**

Example: Factor the binomial. $x^2 - 9$

| Rewrite as $a^2 - b^2$ | $(x)^2 - (3)^2$ |
|---|---|

$$x^2 - 9 = (x-3)(x+3)$$

---

Example: Factor the binomial. $4x^2 - 25$

| Rewrite as $a^2 - b^2$ | $(2x)^2 - (5)^2$ |
|---|---|

$$4x^2 - 25 = (2x-5)(2x+5)$$

---

Example: Factor the binomial. $32x^2 - 18$

| Factor then rewrite as $a^2 - b^2$ | $2(16x^2 - 9)$ <br> $2((4x)^2 - (3)^2)$ |
|---|---|

$$32x^2 - 18 = 2(4x-3)(4x+3)$$

### Factoring The Sum of Two Cubes

$$a^3 + b^3 = (a+b)(a^2 - ab + b^2)$$

Example: Factor the binomial. $8x^3 + 125$

| Rewrite as $a^3 + b^3$ | $(2x)^3 + (5)^3$ |
|---|---|

$$8x^3 + 125 = (2x+5)(4x^2 - 10x + 25)$$

### Factoring The Difference of Two Cubes

$$a^3 - b^3 = (a-b)(a^2 + ab + b^2)$$

Example: Factor the binomial. $27x^3 - 8$

| Rewrite as $a^3 - b^3$ | $(3x)^3 - (2)^3$ |
|---|---|

$$27x^3 - 8 = (3x-2)(9x^2 + 6x + 4)$$

| | **Factoring Polynomials – Useful Formulas** |
|---|---|
| Perfect Square Trinomials | $(a+b)^2 = a^2 + 2ab + b^2$ <br> $(a-b)^2 = a^2 - 2ab + b^2$ |
| Difference of 2 Squares | $a^2 - b^2 = (a-b)(a+b)$ |
| Sum and Diff. of 2 Cubes | $a^3 + b^3 = (a+b)(a^2 - ab + b^2)$ <br> $a^3 - b^3 = (a-b)(a^2 + ab + b^2)$ |
| Distributive Property | $(a+b)(c+d+e)$ <br> $= ac + ad + ae + bc + bd + be$ |
| Quadratic Formula | If: $ax^2 + bx + c = 0$ <br> Then: $x = \dfrac{-b \pm \sqrt{b^2 - 4ac}}{2a}$ |
| Binomial Expansion | $(a+b)^n = \sum_{k=0}^{n} \binom{n}{k} a^{n-k} \cdot b^k$ |
| Combination | $\binom{n}{k} = \dfrac{n!}{k! \cdot (n-k)!}$ |
| Pascal's Triangle | $\begin{array}{ccccccccc} & & & & 1 & & & & \\ & & & 1 & & 1 & & & \\ & & 1 & & 2 & & 1 & & \\ & 1 & & 3 & & 3 & & 1 & \\ 1 & & 4 & & 6 & & 4 & & 1 \\ 1 & 5 & & 10 & & 10 & & 5 & 1 \end{array}$ |

| **Factoring Trinomials:** $ax^2 + bx + c$ |
|---|
| **With $a = 1$   All Positive Terms** |

The goal of factoring trinomials is to rewrite it as the product of two linear factors, in the form:
$$(x - p)(x - q)$$

Example: Factor the trinomial. $x^2 + 6x + 8$

| | |
|---|---|
| Start with this form.<br>Note: $x \cdot x = x^2$ | $(x\quad)(x\quad)$ |
| Find two numbers:<br>• $2 \cdot 4 = 8$<br>• $2 + 4 = 6$ | $(x\quad 2)(x\quad 4)$<br>$(x + 2)(x + 4)$ |
| Check Middle Term:<br><br>• Multiply the <u>inner</u> & <u>outer</u> numbers.<br>• When added, it should equal the middle term ($6x$) | $x^2 + 6x + 8$<br>$(x + 2)(x + 4)$<br>$2x$<br>$+$<br>$4x$<br>$=$<br>$6x$ |

$$x^2 + 6x + 8 = (x + 2)(x + 4)$$

| Factoring Trinomials: $ax^2 + bx + c$ |
|---|
| With $a = 1$     Negative $b$ |

Example: Factor the trinomial. $x^2 - 6x + 8$

| | |
|---|---|
| Start with this form.<br>Note: $x \cdot x = x^2$ | $(x \quad )(x \quad )$ |
| Find two numbers:<br>• $(-2)(-4) = 8$<br>• $-2 - 4 = -6$ | $(x - 2)(x - 4)$ |
| Check Middle Term:<br><br>• Multiply the <u>inner</u> & <u>outer</u> numbers.<br>• When added, it should equal the middle term $(-6x)$ | $x^2 - 6x + 8$<br>$(x - 2)(x - 4)$<br>$-2x$<br>$+$<br>$-4x$<br>$=$<br>$-6x$ |
| $x^2 - 6x + 8 = (x - 2)(x - 4)$ ||

| Factoring Trinomials: $ax^2 + bx + c$ With $a = 1$ — Negative $c$ ||
|---|---|
| Example: Factor the trinomial. $x^2 + 3x - 10$ ||
| Start with this form. Note: $x \cdot x = x^2$ | $(x \quad )(x \quad )$ |
| Find two numbers:<br>• $(-2)(5) = -10$<br>• $-2 + 5 = 3$ | $(x - 2)(x + 5)$ |
| Check Middle Term:<br>• Multiply the <u>inner</u> & <u>outer</u> numbers.<br>• When added, it should equal the middle term $(+3x)$ | $x^2 + 3x - 10$<br>$(x - 2)(x + 5)$<br><br>$-2x$<br>$+$<br>$+5x$<br>$=$<br>$+3x$ |
| $x^2 + 3x - 10 = (x - 2)(x + 5)$ ||

| Factoring Trinomials: With $a = 1$ | $ax^2 + bx + c$ Negative b and c |
|---|---|
| Example: Factor the trinomial. $x^2 - 3x - 10$ ||
| Start with this form. Note: $x \cdot x = x^2$ | $(x \quad)(x \quad)$ |
| Find two numbers: <br> • $(-2)(5) = -10$ <br> • $2 - 5 = -3$ | $(x + 2)(x - 5)$ |
| Check Middle Term: <br><br> • Multiply the inner & outer numbers. <br> • When added, it should equal the middle term $(-3x)$ | $x^2 - 3x - 10$ <br> $(x + 2)(x - 5)$ <br><br> $+ 2x$ <br> $+$ <br> $- 5x$ <br> $=$ <br> $- 3x$ |
| $x^2 - 3x - 10 = (x + 2)(x - 5)$ ||

| Factoring Trinomials: $ax^2 + bx + c$ With $a \neq 1$ ||
|---|---|
| Example: Factor the trinomial. $2x^2 + 8x + 6$ ||
| Note: Common factor<br>Factor out the 2 | $2(x^2 + 4x + 3)$ |
| Start with this form.<br>Note: $x \cdot x = x^2$ | $2(x \quad)(x \quad)$ |
| Find two numbers:<br>• $(1)(3) = 3$ | $2(x + 1)(x + 3)$ |
| Check Middle Term:<br><br>• Multiply the <u>inner</u> & <u>outer</u> numbers.<br>• When added, it should equal the middle term $(+4x)$ | $x^2 + 4x + 3$<br>$(x + 1)(x + 3)$<br>$+ 1x$<br>$+$<br>$+ 3x$<br>$=$<br>$+ 4x$ |
| $2x^2 + 8x + 6 = 2(x + 1)(x + 3)$ ||

| Factoring Trinomials: With $a \neq 1$ | $ax^2 + bx + c$ Negative $c$ |
|---|---|
| Example: Factor the trinomial. $2x^2 + x - 15$ ||
| Start with this form. Note: $2x \cdot x = 2x^2$ | $(2x \quad)(x \quad)$ |
| Find two numbers: <br> • $(-5)(3) = -15$ | $(2x - 5)(x + 3)$ |
| Check Middle Term: <br><br> • Multiply the <u>inner</u> & <u>outer</u> numbers. <br> • When added, it should equal the middle term $(+1x)$ | $x^2 + x - 15$ <br> $(2x - 5)(x + 3)$ <br><br> $-5x$ <br> $+$ <br> $+6x$ <br> $=$ <br> $+x$ |
| $2x^2 + x - 15 = (2x - 5)(x + 3)$ ||

| **Factoring Trinomials:** $ax^2 + bx + c$ |
|---|
| **With** $a \neq 1$     **Negative** $b$ **&** $c$ |

| Example: Factor the trinomial. $3x^2 - 2x - 8$ ||
|---|---|
| Start with this form. <br> Note: $3x \cdot x = 3x^2$ | $(3x \quad)(x \quad)$ |
| Find two numbers: <br> • $(4)(-2) = -8$ | $(3x + 4)(x - 2)$ |
| Check Middle Term: <br><br> • Multiply the <u>inner</u> & <u>outer</u> numbers. <br> • When added, it should equal the middle term $(-2x)$ | $x^2 - 2x - 8$ <br> $(3x + 4)(x - 2)$ <br><br> $+\ 4x$ <br> $+$ <br> $-\ 6x$ <br> $=$ <br> $-\ 2x$ |
| $3x^2 - 2x - 8 = (3x + 4)(x - 2)$ ||

# Why Factor Polynomials

## Why Factor Polynomials?

Often, $y$ is a function of $x$. So, we say: $y = f(x)$
And often, the function is a polynomial.

If $y = f(x)$ and the function is fully factored.
Then, it will be in the form: $y = (x - p)(x - q)$

When the function is fully factored, it is easy to find the values of $x$ that will make the function $= 0$
In the above example, $y = 0$ if $x = p$ or $q$

## Zero Product Principle

If $a \cdot b = 0$     Then, $a = 0$ or $b = 0$

Example: Given $y = f(x) = x(x - 2)(x + 5)$
Find the values of $x$ that make the function $= 0$

Answer: $x = \{0, 2, -5\}$

**Conclusion:** If $y = f(x)$ and $f(x)$ is fully factored. Then, it is easy to find the zeros of the function by using the Zero Product Principle.

# Completing The Square

| | Equations With Perfect Squares Are Easy to Solve |
| --- | --- |

| | |
| --- | --- |
| Example #1 | $x^2 = 25$ $\sqrt{x^2} = \pm\sqrt{25}$ $x = \pm 5$ |
| Check the solutions | $(5)^2 = 25$  OK $(-5)^2 = 25$  OK |
| Conclusion | Equations with perfect squares are easy to solve. Just take the square root of both sides. Add the $\pm$ sign. |

| | |
| --- | --- |
| Example #2 | $(x-3)^2 = 16$ $\sqrt{(x-3)^2} = \pm\sqrt{16}$ $x - 3 = \pm 4$ $x = 3 \pm 4 = 7, -1$ |
| Conclusion | Equations with perfect squares are easy to solve. Remember to add $\pm$ sign. |

## Completing the Square

If a 2nd degree equation $ax^2 + bx + c = n$
We can rearrange it to have a perfect square, making it easier to solve. $(x + c)^2 = m$

The technique is explained with an example, below.

---

Example: Given $x^2 - 6x - 62 = 50$
Solve for $x$, by completing the square.

| | |
|---|---|
| Separate terms with $x$ from the numbers. | $x^2 - 6x = 62 + 50$ <br> $x^2 - 6x = 112$ |
| Take the coefficient of $x$, divide by 2, Then square it. Add this to both sides of eqn. | $x^2 - 6x + \left(\frac{6}{2}\right)^2 = 112 + \left(\frac{6}{2}\right)^2$ <br> $x^2 - 6x + (3)^2 = 112 + (3)^2$ <br> $(x - 3)^2 = 121$ |
| Take the square root of both sides. Remember the $\pm$ sign Solve it. | $\sqrt{(x-3)^2} = \pm\sqrt{121}$ <br> $(x - 3) = \pm 11$ <br> $x = 3 \pm 11$ <br> $x = 14, -8$ |

| Completing the Square – Example 1 ||
|---|---|
| Example: Given $x^2 + 20x - 5 = -65$<br>Solve for $x$, by completing the square. ||
| Separate terms with $x$ from the numbers. | $x^2 - 20x = -65 + 5$<br>$x^2 - 20x = -60$ |
| Divide coefficient of $x$ by 2, then square it. Add to both sides. | $x^2 - 20x + \left(\frac{20}{2}\right)^2 = -60 + \left(\frac{20}{2}\right)^2$<br>$x^2 - 6x + (10)^2 = -60 + (10)^2$<br>$(x - 10)^2 = 40$ |
| Take the square root of both sides. Remember the $\pm$ sign Solve it. | $\sqrt{(x-10)^2} = \pm\sqrt{40}$<br>$(x - 10) = \pm 2\sqrt{10}$<br>$x \approx 10 \pm 6.3$<br>$x \approx 16.3, 3.7$ |

## Completing the Square – Example 2

**Example:** Given $2x^2 - x + 7 = 10$
Solve for $x$, by completing the square.

| | |
|---|---|
| Coeff. of $x^2$ must be one. Divide both sides by 2. | $x^2 - \frac{1}{2}x + \frac{7}{2} = \frac{10}{2}$ |
| Separate terms with $x$ from the numbers. | $x^2 - \frac{1}{2}x = \frac{3}{2}$ |
| Divide coefficient of $x$ by 2, then square it. Add to both sides. | $x^2 - \frac{1}{2}x + \left(\frac{1}{4}\right)^2 = \frac{3}{2} + \left(\frac{1}{4}\right)^2$ <br> $x^2 - \frac{1}{2}x + \left(\frac{1}{4}\right)^2 = \frac{3}{2} + \left(\frac{1}{4}\right)^2$ <br> $\left(x - \frac{1}{4}\right)^2 = \frac{24}{16} + \frac{1}{16} = \frac{25}{16}$ |
| Take the square root of both sides. Remember the $\pm$ sign. Solve it. | $\left(x - \frac{1}{4}\right) = \pm\sqrt{\frac{25}{16}}$ <br> $x = \frac{1}{4} \pm \frac{5}{4}$ <br> $x = \frac{6}{4}, -\frac{4}{4} = \frac{3}{2}, -1$ |

# Quadratic Formula

## Quadratic Formula

If $ax^2 + bx + c = 0$

Then $x = \dfrac{-b \pm \sqrt{b^2 - 4ac}}{2a}$

---

The **Quadratic Formula** is one of the most important tools in your math tool-box. It is used in many math topics involving Quadratic Equations (2nd degree equations). **Memorize it!**

---

$a \neq 0$ for two reasons:
- If $a = 0$ then there is no $x^2$ term so it's a 1st degree equation.
- If $a = 0$ then the denominator, in the formula, is zero. Division by zero is undefined.

---

A little name confusion…
- "Quad" means four, but "Quadratic" comes from the Latin word "quadrum," meaning "to make square." But, don't worry about all of this!
- **Just remember, the Quadratic Formula is used to solve 2nd degree equations.**

## Quadratic Formula – Ex. 01a
### Solve Quadratic Equation (Solution #1)

Given: $x^2 + 2x - 8 = 0$

Solve the equation for $x$. In other words, find the value(s) of $x$ that makes the equation true.

Solution #1: Use the Quadratic Formula

| | |
|---|---|
| Identify the coefficients. | $a = 1, \quad b = 2, \quad c = -8$ |
| Use the Quadratic Formula | $x = \dfrac{-b \pm \sqrt{b^2 - 4ac}}{2a}$ <br><br> $x = \dfrac{-2 \pm \sqrt{(2)^2 - 4(1)(-8)}}{2(1)}$ |
| Solve for $x$ | $x = \dfrac{-2 \pm \sqrt{4 + 32}}{2} = \dfrac{-2 \pm \sqrt{36}}{2}$ <br><br> $x = \dfrac{-2 \pm 6}{2} = \dfrac{-2}{2} \pm \dfrac{6}{2}$ <br><br> $x = -1 \pm 3$ <br><br> $x = 2, -4$ |

## Quadratic Formula – Ex. 01b
## Solve Quadratic Equation (Solution #2)

Given: $x^2 + 2x - 8 = 0$

Solve the equation for $x$. In other words, find the value(s) of $x$ that makes the equation true.

Solution #2: Use factoring.

| | |
|---|---|
| Factor the trinomial | $x^2 + 2x - 8 = 0$<br>$(x-2)(x+4) = 0$ |
| Use the zero product principle.<br><br>If $a \cdot b = 0$<br>Then<br>$a = 0$ or $b = 0$ | If either factor is zero, the left side is equal to zero.<br><br>By observation ...<br>$x = 2, -4$ |
| Conclusions | When solving a quadratic equation, if possible, factor it and just use the zero product principle.<br><br>Just use the quadratic formula if you can't factor it! |

## Quadratic Formula – Ex. 02
## Solve Quadratic Equation

Given: $6x^2 - 7x - 5 = 0$
Solve the equation, using the Quadratic Formula.
Then, use your answer to factor the trinomial.

| | |
|---|---|
| Identify the coefficients. | $a = 6, \; b = -7, \; c = -5$ |
| Use the Quadratic Formula | $x = \dfrac{-b \pm \sqrt{b^2 - 4ac}}{2a}$ <br><br> $x = \dfrac{7 \pm \sqrt{(-7)^2 - 4(6)(-5)}}{2(6)}$ |
| Solve for $x$ | $x = \dfrac{7 \pm \sqrt{49 + 120}}{12} = \dfrac{7 \pm \sqrt{169}}{12}$ <br><br> $x = \dfrac{7 \pm 13}{12} = \dfrac{20}{12}, \dfrac{-6}{12}$ <br><br> $x = \dfrac{5}{3}, \; -\dfrac{1}{2}$ |
| Factor the trinomial | $6x^2 - 7x - 5 = 0$ <br><br> $\left(x - \dfrac{5}{3}\right) \cdot \left(x + \dfrac{1}{2}\right) = 0$ |

## Quadratic Formula – Ex. 03a
## Solve Quadratic Equation   (Solution #1)

Given: $x^2 - 9 = 0$

Solve the equation, using the Quadratic Formula.
Then, solve it again, by factoring.

Solution #1: Use the Quadratic Formula

| | |
|---|---|
| Note: This is a 2nd degree eqn. | $x^2 - 9 = 0$ <br> $x^2 + 0x - 9 = 0$ |
| Identify the coefficients. | $a = 1, \quad b = 0, \quad c = -9$ |
| Use the Quadratic Formula | $x = \dfrac{-b \pm \sqrt{b^2 - 4ac}}{2a}$ <br><br> $x = \dfrac{0 \pm \sqrt{0 - 4(1)(-9)}}{2(1)}$ |
| Solve for $x$ | $x = \dfrac{\pm\sqrt{36}}{2} = \dfrac{\pm 6}{2} = \pm 3$ <br><br> $x = 3, -3$ |

## Quadratic Formula – Ex. 03b
## Solve Quadratic Equation (Solution #2)

Given: $x^2 - 9 = 0$

Solve the equation, using the Quadratic Formula. Then, solve it again, by factoring.

Solution #2: Use factoring.

| | |
|---|---|
| Factor the trinomial<br><br>Note: Difference of two squares | $x^2 - 9 = 0$<br>$x^2 - 3^2 = 0$<br>$(x - 3)(x + 3) = 0$ |
| Use the zero product principle.<br><br>If $a \cdot b = 0$<br>Then<br>$a = 0$ or $b = 0$ | If either factor is zero, the left side is equal to zero.<br><br>By observation ...<br>$x = 3, -3$ |
| Conclusion | When solving a quadratic equation, if possible, factor it and just use the zero product principle. |

## Quadratic Formula – Ex. 04a
## Solve Quadratic Equation   (Solution #1)

Given: $x^2 - 7 = 0$

Solve the equation, using the Quadratic Formula.
Then, solve it again, by factoring.

**Solution #1: Use the Quadratic Formula**

| | |
|---|---|
| Note: This is a 2nd degree eqn. | $x^2 - 7 = 0$ <br> $x^2 + 0x - 7 = 0$ |
| Identify the coefficients. | $a = 1, \quad b = 0, \quad c = -7$ |
| Use the Quadratic Formula | $x = \dfrac{-b \pm \sqrt{b^2 - 4ac}}{2a}$ <br><br> $x = \dfrac{0 \pm \sqrt{0 - 4(1)(-7)}}{2(1)}$ |
| Solve for x | $x = \dfrac{\pm\sqrt{28}}{2} = \dfrac{\pm\sqrt{4 \cdot 7}}{2}$ <br><br> $x = \dfrac{\pm 2\sqrt{7}}{2} = \pm\sqrt{7}$ <br><br> $x = \sqrt{7}, \; -\sqrt{7}$ |

## Quadratic Formula – Ex. 04b
### Solve Quadratic Equation (Solution #2)

Given: $x^2 - 7 = 0$
Solve the equation, using the Quadratic Formula.
Then, solve it again, by factoring.

Solution #2: Use factoring.

| | |
|---|---|
| Factor the trinomial<br><br>Note: Difference of two squares | $x^2 - 7 = 0$<br>$x^2 - (\sqrt{7})^2 = 0$<br>$(x - \sqrt{7})(x + \sqrt{7}) = 0$ |
| Use the zero product principle.<br><br>If $a \cdot b = 0$<br>Then<br>$a = 0$ or $b = 0$ | If either factor is zero, the left side is equal to zero.<br><br>By observation ...<br>$x = \sqrt{7}, -\sqrt{7}$ |
| Conclusion | When solving a quadratic equation, if possible, factor it and just use the zero product principle. |

## Quadratic Formula – Ex. 05
## Solve Quadratic Equation

Given: $4x^2 + 9 = 0$

Solve the equation, using the Quadratic Formula. Then, use your answer to factor the trinomial.

Note: Can't factor using the difference of two squares because it's the sum of two squares!

| | |
|---|---|
| Identify the coefficients. | $a = 4, \quad b = 0, \quad c = 9$ |
| Use the Quadratic Formula | $x = \dfrac{-b \pm \sqrt{b^2 - 4ac}}{2a}$ <br><br> $x = \dfrac{0 \pm \sqrt{0 - 4(4)(9)}}{2(4)}$ |
| Solve for x | $x = \dfrac{\pm\sqrt{0 - 144}}{8} = \dfrac{\pm\sqrt{-1}\sqrt{144}}{8}$ <br><br> $x = \dfrac{\pm 12i}{8} = \dfrac{\pm 3i}{2} = \dfrac{3i}{2}, -\dfrac{3i}{2}$ |
| Factor the trinomial | $\left(x - \dfrac{3i}{2}\right) \cdot \left(x + \dfrac{3i}{2}\right) = 0$ <br><br> $\dfrac{1}{2}(2x - 3i) \cdot \dfrac{1}{2}(2x + 3i) = 0$ <br><br> $\dfrac{1}{4}(2x - 3i)(2x + 3i) = 0$ |

# Function Composition

## Function Composition

Function composition is an operation involving two functions ($g$ and $h$) to produce a 3$^{rd}$ function ($f$).
$$f = g \circ h$$

The composition of two functions, $g$ and $h$ has several notations, as shown below:

| | |
|---|---|
| $g \circ h$ | Pronounced "$g$ of $h$" or "$g$ circle $h$" <br> Here, output from $h$ is input to $g$ |
| $g(h(x))$ | Pronounced "$g$ of $h$ of $x$" <br> Here, $x$ is input to $h$ <br> The output from $h(x)$ is the input to $g$ |

---

In general, $g(h(x)) \neq h(g(x))$

But if
$$g(h(x)) = h(g(x)) = x$$
then they are inverse functions.

## Function Composition -- Example Set 1

Given: $g(x) = 3x$ and $h(x) = \sqrt{x}$
Find the following.

| | |
|---|---|
| $g(h(4))$ | $= g(\sqrt{4}) = g(2) = 3(2) = 6$ |
| $h(g(4))$ | $= h(3(4)) = h(12) = \sqrt{12} = 2\sqrt{3}$ |
| $g(h(9))$ | $= g(\sqrt{9}) = g(3) = 3(3) = 9$ |
| $h(g(9))$ | $= h(3(9)) = h(27) = \sqrt{27} = 3\sqrt{3}$ |
| $g(h(x))$ | $= g(\sqrt{x}) = 3\sqrt{x}$ |
| $h(g(x))$ | $= h(3x) = \sqrt{3x}$ |
| $h \circ g$ | $= h(g(x)) = h(3x) = \sqrt{3x}$ |

## Function Composition -- Example Set 2

Given graphs of $g$ and $h$, find the indicated values.

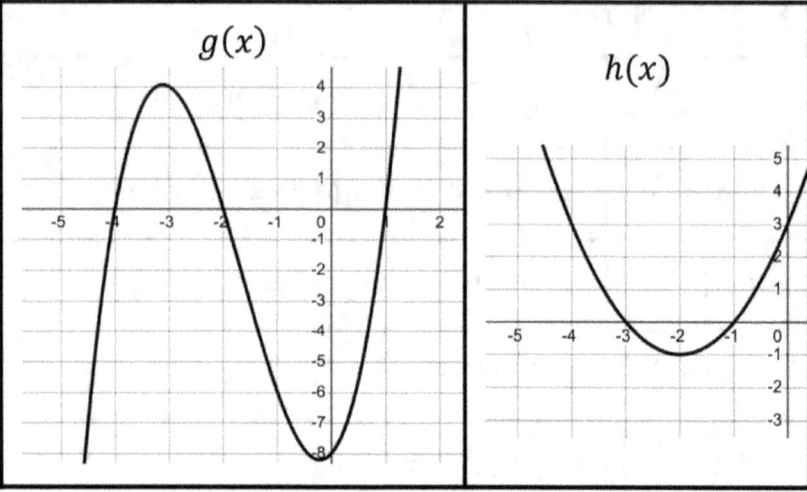

| | | |
|---|---|---|
| $g(h(-3))$ | $= g(0)$ | $= -8$ |
| $h(g(-4))$ | $= h(0) = 3$ | |
| $g(h(-2))$ | $= g(-1) = -6$ | |
| $g(h(-0.5))$ | $= g(1) = 0$ | |

## Function Composition -- Example Set 3

Given values of $g$ and $h$ in the table below, find the indicated values.

| $x \rightarrow$ | 0 | 1 | 2 | 3 | 4 | 5 | 6 |
|---|---|---|---|---|---|---|---|
| $g(x)$ | 0 | 2 | 4 | 6 | 8 | 10 | 12 |
| $h(x)$ | 5 | 3 | 1 | 0 | 4 | 6 | 2 |

| | |
|---|---|
| $g(h(0))$ | $= g(5) = 10$ |
| $h(g(0))$ | $= h(0) = 5$ |
| $g(h(1))$ | $= g(3) = 6$ |
| $h(g(1))$ | $= h(2) = 1$ |
| $g(h(2))$ | $= g(1) = 2$ |
| $h(g(2))$ | $= h(4) = 4$ |
| $g(h(3))$ | $= g(0) = 0$ |
| $h(g(3))$ | $= h(6) = 2$ |

# Inverse Functions

| Inverse of a Function |
|---|
| Let $f(x)$ and $g(x)$ be two **one-to-one** functions. <br> If $\quad f \circ g \;=\; g \circ f \;=\; x$ <br> Then $\quad f(x)$ and $g(x)$ are inverses of each other. |

| Notation: | $f^{-1}(x) \;=\;$ The inverse of $f(x)$ <br> $g^{-1}(x) \;=\;$ The inverse of $g(x)$ |
|---|---|

| One-to-One Functions |
|---|
| A function is **one-to-one** if there is no horizontal line that intersects its graph more than once. <br><br> In other words … A function is **one-to-one** if it passes the horizontal line test. |

| $y = x^3$ <br> Is one-to-one | 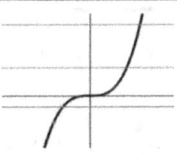 |
|---|---|
| $y = x^2$ <br> Is NOT one-to-one | 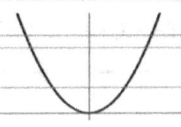 |

| How to Find the Inverse of a Function |
|---|
| Basically, we just switch the x's and y's  The simple example below, demonstrates the process. |

| Inverse Function – Example 0 ||
|---|---|
| Given: $f(x) = 2x - 5$    Find: $f^{-1}(x)$ ||
| Rewrite function as $y = f(x)$ | $y = 2x - 5$ |
| Switch the x's and y's | $x = 2y - 5$ |
| Solve for y | $x = 2y - 5$  $x + 5 = 2y$  $y = \frac{x+5}{2}$ |
| Rename it to $f^{-1}(x)$   We can't call both functions " y " | $f^{-1}(x) = \frac{x+5}{2}$ |

## How to Determine if Two Functions Are Inverses of Each Other

There are two ways to determine if two functions are inverses of each other – Algebraically and Graphically.

Given: $f = 2x - 5$ and $g = \dfrac{x+5}{2}$

| Algebraic Test<br><br>$f \circ g = x$<br>$g \circ f = x$ | $f \circ g = f(g(x))$<br>$\quad = g\left(\dfrac{x+5}{2}\right)$<br>$\quad = 2\left(\dfrac{x+5}{2}\right) - 5 \;= x$ |
|---|---|
| | $g \circ f = g(f(x))$<br>$\quad = g(2x - 5)$<br>$\quad = \dfrac{(2x-5)+5}{2} \;= x$ |
| Graphical Test<br><br>Graphs of the two functions should be reflections over the $y = x$ axis | 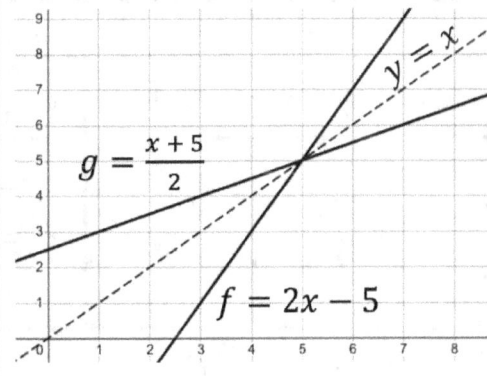 |

## Inverse of a Function – Example Set 1

| | |
|---|---|
| Given: $f(x) = x^3 + 7$ | $y = x^3 + 7$ |
| Switch $x$ & $y$<br>Solve for $y$ | $x = y^3 + 7$<br>$y^3 = x - 7$<br>$y = \sqrt[3]{x - 7}$ |
| Inverse of $f(x)$ | $f^{-1}(x) = \sqrt[3]{x - 7}$ |

| | |
|---|---|
| Given: $f(x) = \dfrac{x+3}{5}$ | $y = \dfrac{x+3}{5}$ |
| Switch $x$ & $y$<br>Solve for $y$ | $x = \dfrac{y+3}{5}$<br>$5x = y + 3$<br>$y = 5x - 3$ |
| Inverse of $f(x)$ | $f^{-1}(x) = 5x - 3$ |

| | |
|---|---|
| Given: $f(x) = \sqrt[3]{x+9}$ | $y = \sqrt[3]{x+9}$ |
| Switch $x$ & $y$<br>Solve for $y$ | $x = \sqrt[3]{y+9}$<br>$x^3 = y + 9$<br>$y = x^3 - 9$ |
| Inverse of $f(x)$ | $f^{-1}(x) = x^3 - 9$ |

## Inverse of a Function – Example 2

| | |
|---|---|
| Given: | $f(x) = x^2 + 3$ ; $x \geq 0$ |

| | |
|---|---|
| Graph it | 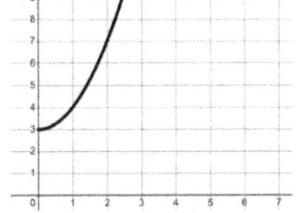 |
| Domain & Range In interval Notation | $D: [0, \infty)$     $R: [3, \infty)$ |
| Write an equation for $f^{-1}(x)$ | $y = x^2 + 3$ ; $x \geq 0$ <br> $x = y^2 + 3$ ; $y \geq 0$ <br> $y = \sqrt{x - 3}$ <br><br> $f^{-1}(x) = \sqrt{x - 3}$ |
| Graph $y = f(x)$ And $y = f^{-1}(x)$ On the same coordinate system. | 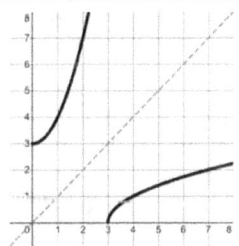 |

Note: Domains and Ranges are opposite (switched) For a function and its' inverse.

# Transformations

## Transformations

Transformations change a given function. Consider the function: $f(x) = x^2$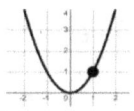

### Horizontal Transformations (Very Nice)

| | | |
|---|---|---|
| $y = f(x) + a$ | V. Shift Up | ↑ |
| $y = a * f(x)$ | V. Stretch | ↕ |
| $y = -f(x)$ | V. Rotation | ↻ |

### Horizontal Transformations (Horrible)

| | | |
|---|---|---|
| $y = f(x + a)$ | H. Shift Left | ← |
| $y = f(a * x)$ | H. Compression | →← |
| $y = f(-x)$ | H. Rotation | ↷ |

## Transformations – Ex. 1a

Given: $y = (x+3)^2 - 5$
Find:   Parent function, transformations, and graph.

| | | |
|---|---|---|
| Parent Function $y = x^2$ | 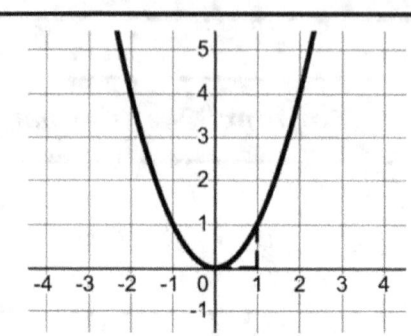 | |
| Vertical and Horizontal Transformations | • Horizontal shift left by 3<br>• Vertical shift down by 5<br>• No stretching or compression | |
| Graph $y = (x+3)^2 - 5$ | 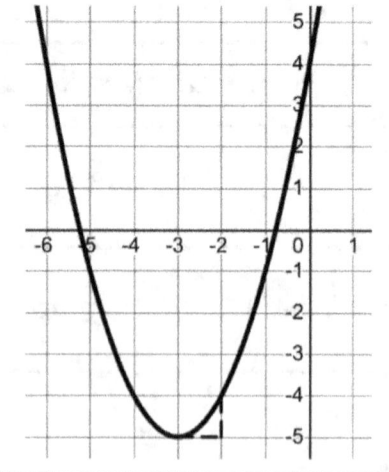 | |

| | |
|---|---|
| **Transformations – Ex. 1b (With T table)** ||
| Given: $y = (x+3)^2 - 5$<br>Find: Parent function, transformations, and graph. ||
| Parent Function $y = x^2$ | 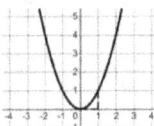 |
| Vertical and Horizontal Transformations | • Horizontal shift left by 3<br>• Vertical shift down by 5<br>• No stretching or compression |
| Use transformations to create extended "T" table. | $x-3$ \| $x$ \| $y$ \| $y-5$<br>$-3$ \| $0$ \| $0$ \| $-5$<br>$-2$ \| $1$ \| $1$ \| $-4$<br>$-1$ \| $2$ \| $4$ \| $-1$ |
| Use points from table to help graph<br><br>$y = (x+3)^2 - 5$ | 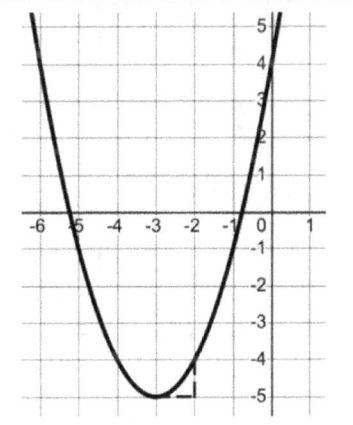 |

## Transformations – Ex. 2a

Given: $y = -\sqrt{2x+6}$
Find: Parent function, transformations, and graph.

| | |
|---|---|
| Parent Function<br><br>$y = \sqrt{x}$ | 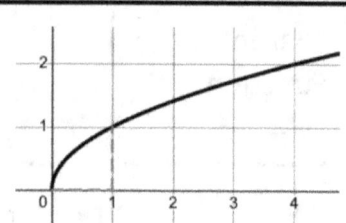 |
| Rewrite eqn. so coefficient of $x$ is 1 | $y = -\sqrt{2x+6}$<br>$y = -\sqrt{2(x+3)}$ |
| Vertical and Horizontal Transformations | • H. Shift left by 3<br>• H. Compression by ½<br>• V. Rotation over $x$ axis |
| Graph<br><br>$y = -\sqrt{2x+6}$ | 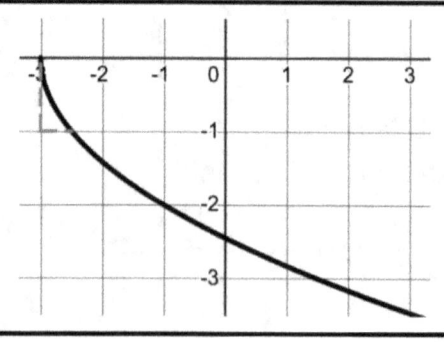 |

# Transformations – Ex. 2b (With T table)

Given: $y = -\sqrt{2x+6}$
Find: Parent function, transformations, and graph.

| | |
|---|---|
| Parent Function $y = \sqrt{x}$ | 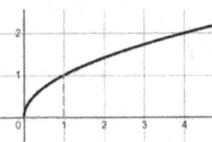 |
| Rewrite eqn. so coeff. of $x$ is 1 | $y = -\sqrt{2x+6}$<br>$y = -\sqrt{2(x+3)}$ |
| Vertical and Horizontal Transformations | • H. Compression by ½<br>• H. Shift left by 3<br>• V. Rotation over $x$ axis |

Use transformations to create extended T-table

| $\frac{1}{2}x - 3$ | $x$ | $y$ | $-y$ |
|---|---|---|---|
| $-3$ | 0 | 0 | 0 |
| $-2.5$ | 1 | 1 | $-1$ |
| $-1$ | 2 | 4 | $-2$ |

Use points from T table to help graph

$y = -\sqrt{2x+6}$

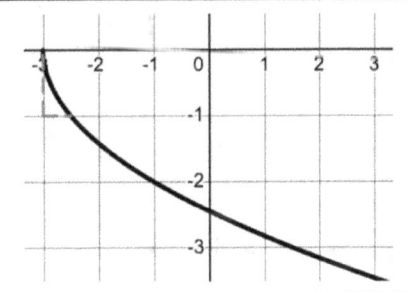

## Transformations – Ex. 3a

Given: $y = -3|x-2|$
Find:    Parent function, transformations, and graph.

| | |
|---|---|
| Parent Function $y = |x|$ | 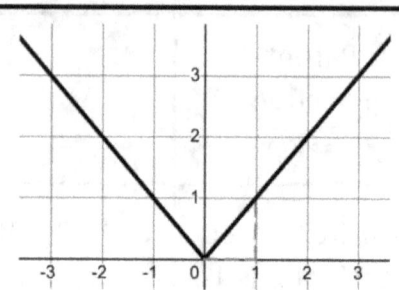 |
| Vertical and Horizontal Transformations | • H. Shift right by 2<br>• V. Stretch by 3<br>• V. Rotation over $x$ axis |
| Graph $y = -3|x-2|$ | 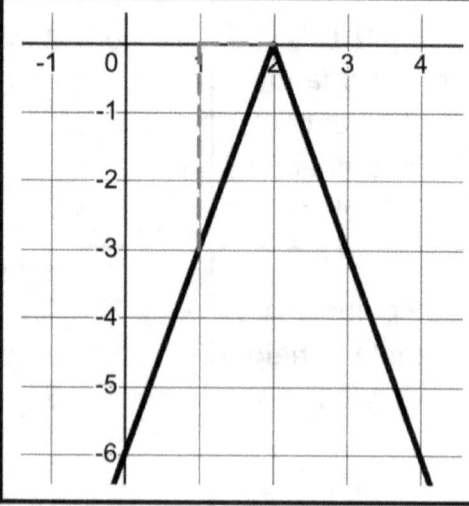 |

## Transformations — Ex. 3b (With T table)

Given: $y = -3|x-2|$
Find: Parent function, transformations, and graph.

| Parent Function $y = |x|$ | 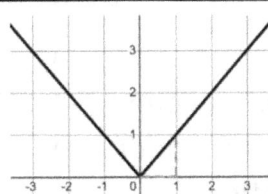 |
|---|---|
| Vertical and Horizontal Transformations | • H. Shift right by 2<br>• V. Stretch by 3<br>• V. Rotation over $x$ axis |

Use transformations to create extended T-table

| $x+2$ | $x$ | $y$ | $-3y$ |
|---|---|---|---|
| 2 | 0 | 0 | 0 |
| 3 | 1 | 1 | -3 |
| 1 | -1 | 1 | -3 |

Use points from T-table to help graph

$y = -3|x-2|$

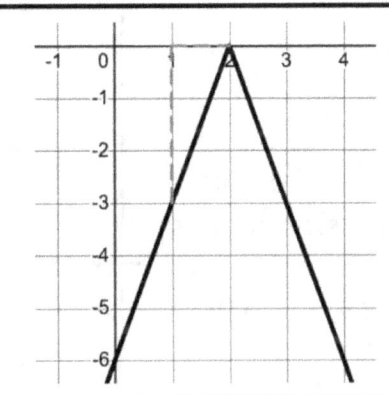

## Transformations – Ex. 4

| | |
|---|---|
| Given: The graph.<br><br>Find:<br>• Parent function,<br>• Transformations<br>• Equation | 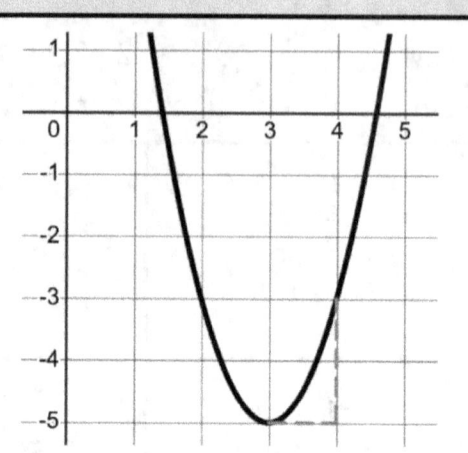 |

| | |
|---|---|
| Parent Function<br>$y = x^2$ | 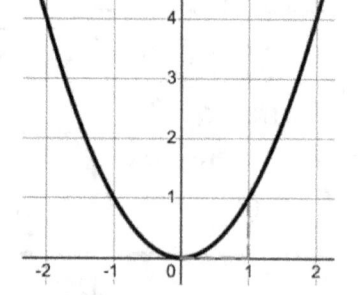 |
| Vertical and Horizontal Transformations | • H. Shift right by 3<br>• V. Shift down by 5<br>• V. Stretch by 2 |
| Equation of Given graph | $y = 2(x - 3)^2 - 5$ |

## Transformations – Ex. 5

| | |
|---|---|
| Given: The graph.<br><br>Find:<br>• Parent function,<br>• Transformations<br>• Equation | 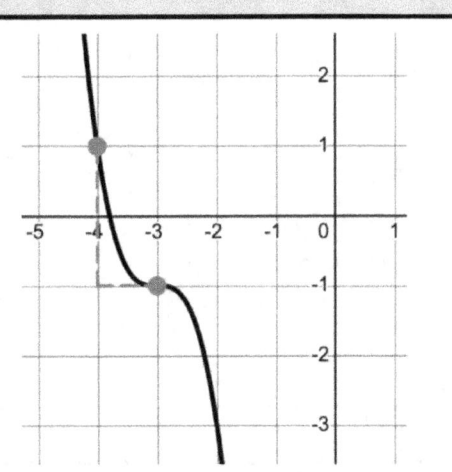 |
| Parent Function<br>$y = x^3$ | 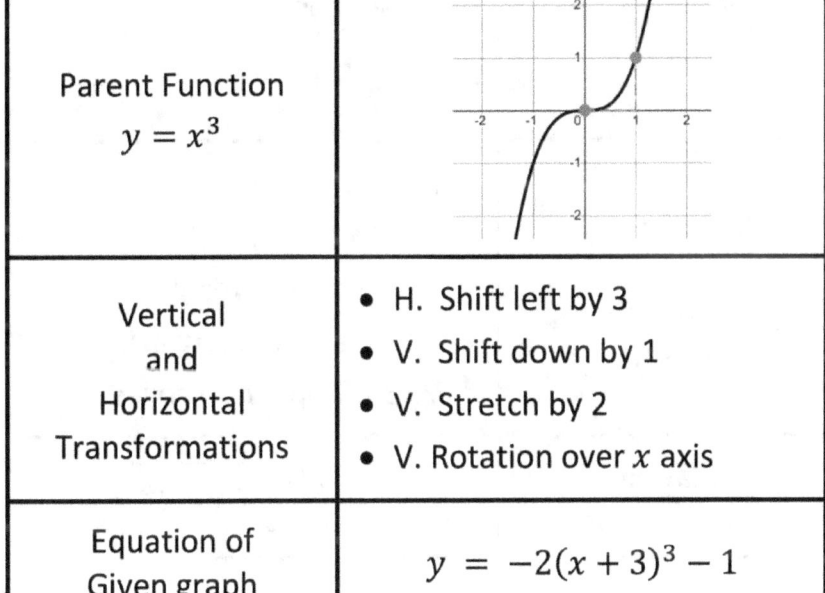 |
| Vertical and Horizontal Transformations | • H. Shift left by 3<br>• V. Shift down by 1<br>• V. Stretch by 2<br>• V. Rotation over $x$ axis |
| Equation of Given graph | $y = -2(x+3)^3 - 1$ |

## Transformations – Ex. 6

| | |
|---|---|
| Given: The graph.<br><br>Find:<br>• Parent function,<br>• Transformations<br>• Equation | 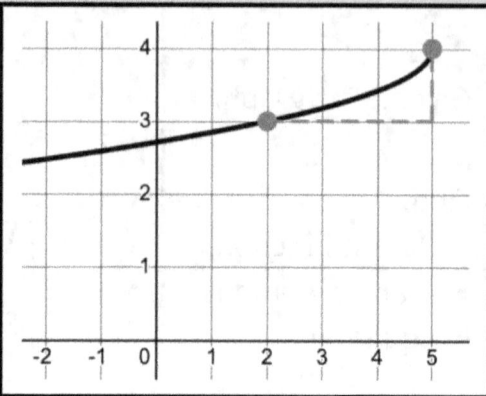 |
| Parent Function<br>$y = \sqrt{x}$ | 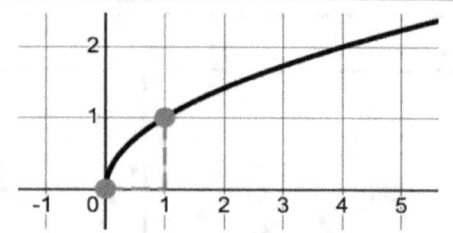 |
| Vertical<br>and<br>Horizontal<br>Transformations | • H. Shift right by 5<br>• H. Stretch left by 3<br>• H. Rotation over $y$ axis<br>• V. Rotation over $x$ axis<br>• V. Shift up by 4 (do last) |
| Equation of<br>Given graph | $y = -\sqrt{-\frac{1}{3}(x-5)} + 4$ |

## Absolute Value Transformations

Taking the absolute value of the **output** is different than taking the absolute value of the **input.**

| | |
|---|---|
| $y = \|f(x)\|$ | Absolute value of the output.<br><br>If you take the absolute value of the output, then all output is positive. All " $y$ " values are positive. |

| | |
|---|---|
| $y = f(\|x\|)$ | Absolute value of the input.<br><br>Here, the " $x$ " values can still be negative or positive. They just act like positive $x$ input.<br><br>Graph has $y$ axis symmetry. |

## Transformations – Ex. 1
### Absolute Value of the Output: $y = |f(x)|$

Given: A set of $(x, y)$ points

Answer the questions below.

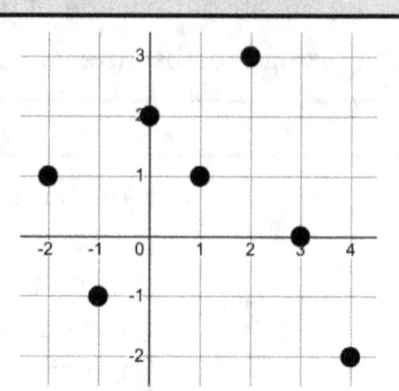

| | |
|---|---|
| Find the domain & range of original set of $(x, y)$ points. | D: $\{-2, -1, 0, 1, 2, 3, 4\}$<br>R: $\{-2, -1, 0, 1, 2, 3\}$ |
| Find the domain & range of $(x, |y|)$ points. | D: $\{-2, -1, 0, 1, 2, 3, 4\}$<br>R: $\{0, 1, 2, 3\}$ |
| Plot the set of points $(x, |y|)$<br><br>Note: All $y$ values are positive | 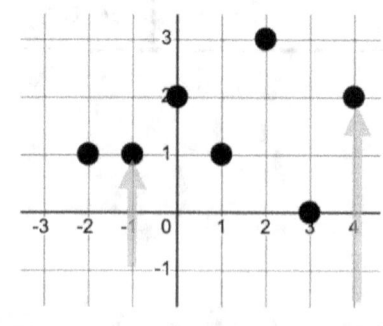 |

## Transformations – Ex. 2
### Absolute Value of the Output: $y = |f(x)|$

| Given: $y = (x-1)^2 - 2$  Answer the questions below. | 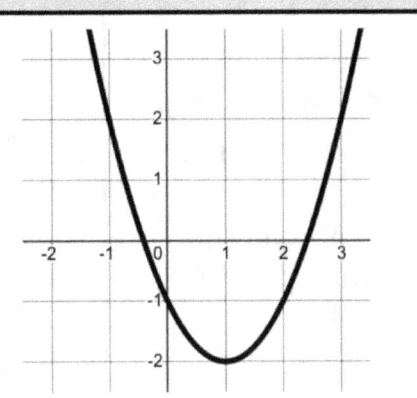 |
|---|---|
| Find the domain & range of original function. | $D: (-\infty, \infty)$  $R: [-2, \infty)$ |
| Find the domain & range of $y = |(x-1)^2 - 2|$ | $D: (-\infty, \infty)$  $R: [0, \infty)$ |
| Plot the function $y = |(x-1)^2 - 2|$  Note: All $y$ values are positive | 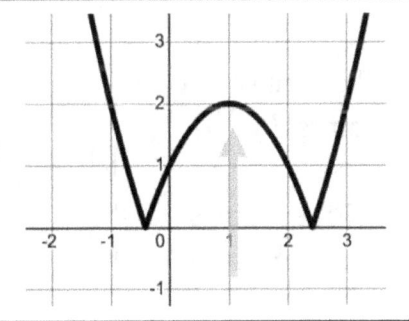 |

## Transformations – Ex. 3
### Absolute Value of the Input:  $y = f(|x|)$

| | |
|---|---|
| Given : A set of $(x, y)$ points<br><br>Answer the questions below. | 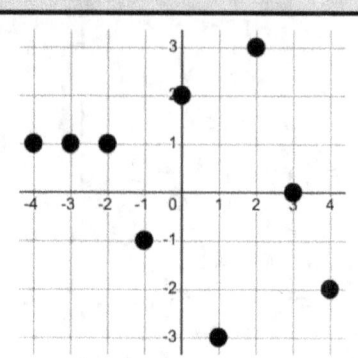 |

| | |
|---|---|
| Find the domain & range of original set of $(x, y)$ points. | $D: \{-4, -3, -2, -1, 0, 1, 2, 3, 4\}$<br>$R: \{-3, -2, -1, 0, 1, 2, 3\}$ |
| Find the domain & range of $(|x|, y)$ points. | $D: \{-4, -3, -2, -1, 0, 1, 2, 3, 4\}$<br>$R: \{-3, -2, 0, 1, 3\}$ |
| Plot the set of points $(x, |y|)$<br><br>Note: $y$-axis symmetry | 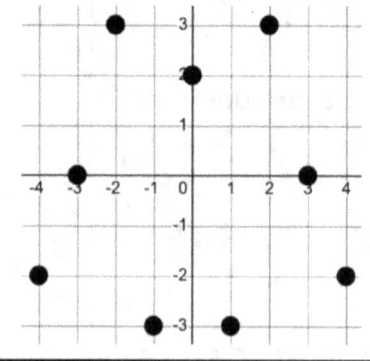 |

## Transformations – Ex. 4
### Absolute Value of the Input: $y = f(|x|)$

Given:

$y = (x - 1)^2 - 2$

Answer the questions below.

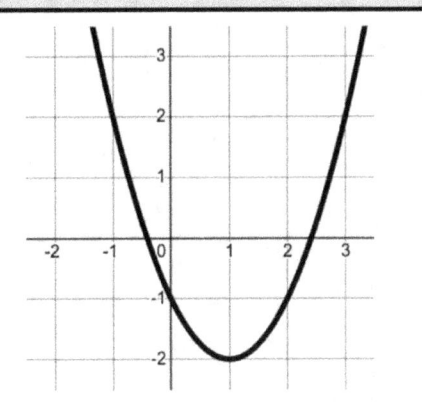

| Find the domain & range of original function. | D: $(-\infty, \infty)$<br>R: $[-2, \infty)$ |
|---|---|
| Find the domain & range of<br>$y = (|x| - 1)^2 - 2$ | D: $(-\infty, \infty)$<br>R: $[-2, \infty)$ |
| Plot the function<br>$y = (|x| - 1)^2 - 2$<br><br>Note: $y$-axis symmetry | 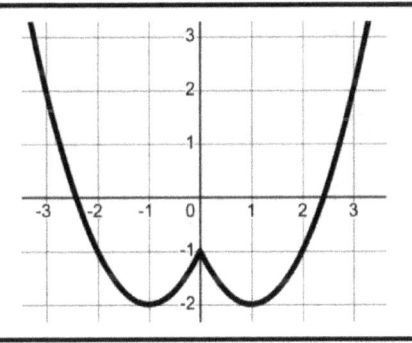 |

# Graphing Quadratic Functions

| Graphing Quadratic Functions |
|---|
| Quadratic Functions may be in one of three forms. |

| Quadratic Equations -- Vertex ||| 
|---|---|---|
| Form | Equation | Vertex $(h, k)$ |
| Standard Form | $f(x) = ax^2 + bx + c$ | $h = -\frac{b}{2a}$ $k = f(h)$ |
| Vertex Form | $f(x) = a(x - h)^2 + k$ | See eqn. |
| Factored Form | $f(x) = a(x - z_1)(x - z_2)$ | $h = \frac{z_1 + z_2}{2}$ $k = f(h)$ |

| Quadratic Equations – Far End Behavior |||
|---|---|---|
| $a =$ Positive | Happy Parabola (opens up) | ☺ |
| $a =$ Negative | Sad Parabola (opens down) | ☹ |

| Graphing Quadratic Functions |
|---|
| Set the quadratic function $= 0$ to find the zeros (or $x$-intercepts). Use the Quadratic Formula to solve it. |

| Quadratic Formula |
|---|
| If $\quad ax^2 + bx + c = 0$ <br> Then $\quad x = \dfrac{-b \pm \sqrt{b^2 - 4ac}}{2a}$ <br> Discriminant $= b^2 - 4ac$ |

| $b^2 - 4ac$ | Zeros | Examples ||
|---|---|---|---|
| 0 | 1 Real | | |
| Positive | 2 Real | | |
| Negative | 0 Real | | |

## Graphing Quadratic Functions – Ex. 1

Given: $f(x) = x^2 + 2x - 15$.  ( $2^{nd}$ degree eqn.)
Find:  Vertex, Zeros, Graph, Domain, Range.

Note: 2 points needed to graph $1^{st}$ degree equation.
    3 points needed to graph $2^{nd}$ degree equation.

| | |
|---|---|
| Vertex | $h = -\dfrac{b}{2a} = -\dfrac{2}{2(1)} = -1$ <br><br> $k = f(-1) = (-1)^2 + 2(-1) - 15$ <br><br> $k = 1 - 2 - 15 = -16$ <br><br> Vertex: $(h, k) = (-1, -16)$ |
| Zeros | $x^2 + 2x - 15 = 0$ <br><br> $(x - 3)(x + 5) = 0 \quad \rightarrow \quad x = 3, -5$ <br><br> Zeros: $(3, 0), (-5, 0)$ |
| Notes | 3 Points: $(-1, -16), (3, 0), (-5, 0)$ <br> Positive $a$: Opens up  (Happy Parabola) |
| Graph | $D: (-\infty, \infty)$ <br> $R: [-16, \infty)$ |

## Graphing Quadratic Functions – Ex. 2

Given: $f(x) = x^2 - 2x - 5$  ( 2nd degree eqn.)
Find:   Vertex, Zeros, Graph, Domain, Range.

| | |
|---|---|
| Vertex | $h = -\dfrac{b}{2a} = \dfrac{2}{2(1)} = 1$ <br> $k = f(1) = (1)^2 - 2(1) - 5 = -6$ <br> Vertex: $(h, k) = (1, -6)$ |
| Zeros | $x^2 - 2x - 5 = 0$   Use Quadratic Formula <br> $x = \dfrac{-b \pm \sqrt{b^2 - 4ac}}{2a}$ <br> $x = \dfrac{2 \pm \sqrt{4 - 4(1)(-5)}}{2(1)} = \dfrac{2 \pm \sqrt{24}}{2} = \dfrac{2 \pm 2\sqrt{6}}{2}$ <br> $x = 1 \pm \sqrt{6}$ <br> Zeros: $(1 + \sqrt{6}, 0), \ (1 - \sqrt{6}, 0)$ |
| Notes | 3 Points: Two zeros and vertex (see above) <br> Positive $a$: Opens up  (Happy Parabola) |
| Graph | 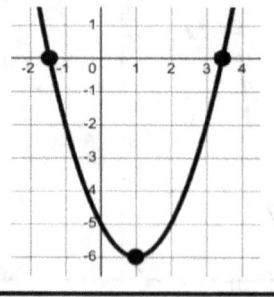  $D: (-\infty, \infty)$ <br> $R: [-6, \infty)$ |

## Graphing Quadratic Functions – Ex. 3

Given: $f(x) = -x^2 + 10x - 25$     ( 2$^{nd}$ degree eqn.)
Find:    Vertex, Zeros, Graph, Domain, Range.

| | |
|---|---|
| Vertex | $h = -\dfrac{b}{2a} = \dfrac{-10}{2(-1)} = 5$ <br><br> $k = f(5) = -(5)^2 + 10(5) - 25 = 0$ <br><br> Vertex: $(h, k) = (5, 0)$ |
| Zeros | $-x^2 + 10x - 25 = 0$    Use Quad. Formula <br><br> $x = \dfrac{-b \pm \sqrt{b^2 - 4ac}}{2a}$ <br><br> $x = \dfrac{-10 \pm \sqrt{100 - 4(-1)(-25)}}{2(-1)} = \dfrac{-10 \pm \sqrt{0}}{-2}$ <br><br> $x = 5$    →    One Zero: $(5, 0)$ |
| Notes | Vertex: $(5, 0)$,    Negative $a$: Opens down <br> Find more points: $f(0) = -25$ <br> $f(1) = -1 + 10 - 25 = -16$ <br> 3 Points: $(5, 0), (0, -25), (1, -16)$ |
| Graph | $D: (-\infty, \infty)$ <br> $R: (-\infty, 0]$ |

| **Graphing Quadratic Functions – Ex. 4** | |
|---|---|
| Given the graph.<br><br>Find the equation. | |
| | |
| Use Vertex Form<br>With<br>$(h, k) = (3, -4)$ | $y = a(x - h)^2 + k$<br>$y = a(x - 3)^2 - 4$ |
| Use another point<br>$(x, y) = (1, 4)$<br>to find " $a$ " | $y = a(x - 3)^2 - 4$<br>$(1, 4) \rightarrow 4 = a(1 - 3)^2 - 4$<br>$4 = a(-2)^2 - 4$<br>$4 = a(4) - 4$<br>$8 = a(4)$<br>$2 = a$ |
| Equation | $y = 2(x - 3)^2 - 4$ |

## Graphing Quadratic Functions – Ex. 5

| | |
|---|---|
| Given graph.<br><br>Find equation. | 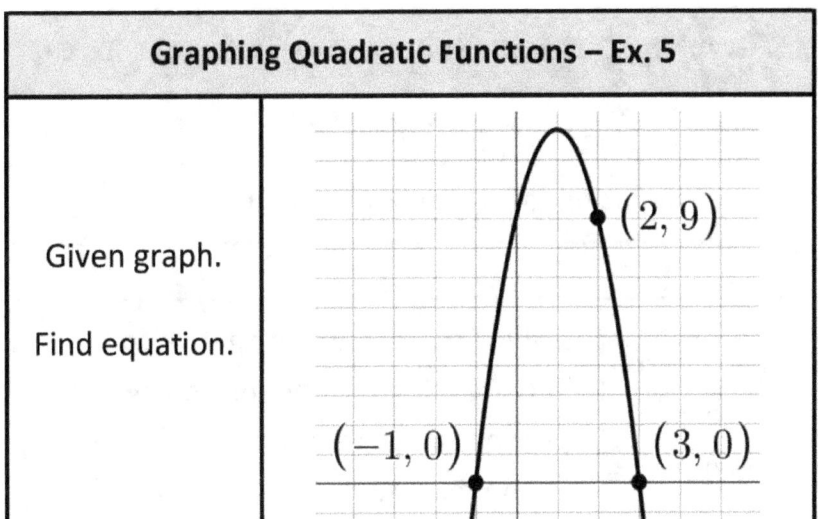 |

| | |
|---|---|
| Use Factored Form<br>With<br>$z_1 = -1$<br>$z_2 = 3$ | $y = a(x - z_1)(x - z_2)$<br><br>$y = a(x + 1)(x - 3)$ |
| Use another point<br>$(x, y) = (2, 9)$<br>to find " $a$ " | $y = a(x + 1)(x - 3)$<br>$(2, 9) \rightarrow 9 = a(2 + 1)(2 - 3)$<br>$9 = a(3)(-1)$<br>$9 = a(-3)$<br>$-3 = a$    (Sad Parabola) |
| Equation | $y = -3(x + 1)(x - 3)$ |
| Note: Equation could be expanded to Standard Form ||

## Graphing Quadratic Functions (Maximum) – Ex. 6

Given: $f(x) = -3(x+1)(x-3)$

Find: The maximum or minimum value.

| | |
|---|---|
| Maximum or Minimum? | Leading coefficient is negative so it is a "sad" parabola. It opens downward. So, it has a maximum and no minimum. |
| Maximum at vertex. | Maximum occurs at the vertex, on the axis of symmetry (AOS). AOS is between the two $x$-intercepts or between the two zeros ($z_1$ and $z_2$). |
| Find the zeros. | Since the given function is in factored form, the zeros are easy to find. $\quad z_1 = -1 \quad\quad z_2 = 3$ |
| Find the vertex. | Vertex at $x = \frac{z_1 + z_2}{2} = \frac{-1+3}{2} = 1$ $y = f(1) = -3(1+1)(1-3) = 12$ Vertex is at $(x, y) = (1, 12)$ |
| Maximum | Maximum value of the function = 12 |

## Graphing Quadratic Functions (Minimum) – Ex. 7

Given: $f(x) = 3x^2 + 9x - 5$

Find: The maximum or minimum value.

| | |
|---|---|
| Maximum or Minimum? | Leading coefficient is positive so it is a "happy" parabola. Opens upward. So, it has a minimum but no max. |
| Find the vertex. | Minimum occurs at the vertex, where $x = -\frac{b}{2a} = -\frac{9}{2(3)} = -\frac{3}{2}$ $y = 3\left(-\frac{3}{2}\right)^2 + 9\left(-\frac{3}{2}\right) - 5 = -\frac{47}{4}$ Vertex at $(x, y) = \left(-\frac{3}{2}, -\frac{47}{4}\right)$ |
| Minimum | Minimum value of the function $= -\frac{47}{4}$ |
| Graph | (−1.5, −11.75) |

## Graphing Quadratic Functions (Intercepts) – Ex. 8a

Given: $f(x) = x^2 + 3x - 4$

Find: All intercepts and vertex. Then, graph it.

| | | |
|---|---|---|
| x intercepts occur when $y = 0$ | $x^2 + 3x - 4 = 0$ <br> $(x - 1)(x + 4) = 0$ <br> $x = 1, -4$ | |
| y intercept occurs when $x = 0$ | $f(0) = 0^2 + 3(0) - 4 =. -4$ <br> y-intercept at $(x, y) = (0, -4)$ | |
| Vertex | AOS between the x intercepts <br><br> $h = \frac{1 + (-4)}{2} = -\frac{3}{2}$ <br><br> $k = f\left(-\frac{3}{2}\right) = \left(-\frac{3}{2}\right)^2 + 3\left(-\frac{3}{2}\right) - 4$ <br><br> $k = \frac{9}{4} - \frac{9}{2} - 4$ <br><br> $k = \frac{9}{4} - \frac{18}{4} - \frac{16}{4} = -\frac{25}{4} = -6.25$ <br><br> Vertex at $(h, k) = (-1.5, -6.25)$ | |

## Graphing Quadratic Functions (Intercepts) – Ex. 8b

Given: $f(x) = x^2 + 3x - 4$

Find: All intercepts and vertex. Then, graph it.

| | |
|---|---|
| Previously Found | $x$ intercepts: $x = 1, -4$ <br> $y$ intercepts: $y = -4$ <br> Vertex: $(h, k) = (-1.5, -6.25)$ |
| Graph | 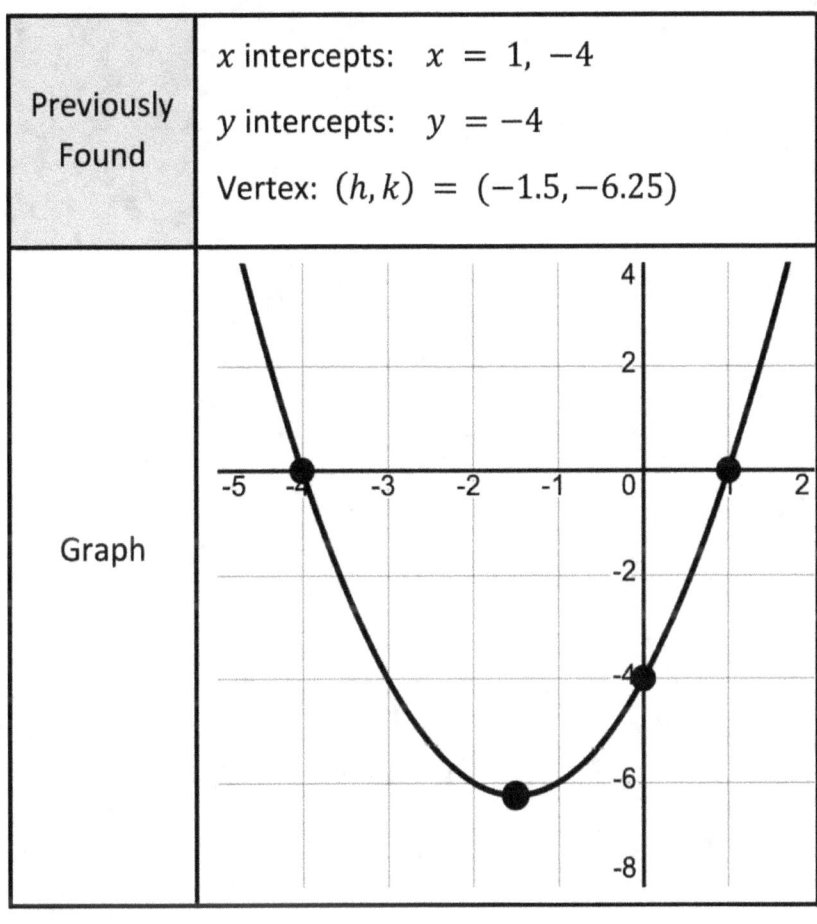 |

# Graphing Polynomial Functions

## Graphing Polynomial Functions

In the previous section, we graphed quadratic functions, which are $2^{nd}$ degree functions.
In this section, we will graph polynomial functions, which are functions with degrees $\geq 1$.

Polynomial functions in factored form:
$$y = a(x - z_1)^i (x - z_2)^j \ldots (x - z_n)^k$$
Degree = Sum of exponents = $i + j + \cdots + k$

## Polynomial Functions – Far End Behavior

| Far-end behavior depends on the degree | Similar to | | Positive $a$ | Negative $a$ |
|---|---|---|---|---|
| | Odd Degree | $1^{st}$ Degree | ↙ ↗ | ↖ ↘ |
| | Even Degree | $2^{nd}$ Degree | ↖ ↗ | ↙ ↘ |

Odd Degree Positive "$a$"

| Turning Points | Number of times a polynomial function changes direction (or wiggles around). Number of turning points = Degree $-1$ |
|---|---|

## Graphing Polynomial Functions (Continued)

The zeros (or roots) of a polynomial function are easy to find if the polynomial is in factored form:

$$y = a(x - z_1)^i (x - z_2)^j \ldots (x - z_n)^k$$

Degree = Sum of exponents = $i + j + \cdots + k$

## Polynomial Functions – Behavior at Zeros

| Behavior at zeros depends on multiplicity. | The multiplicity of a zero is the number of times it occurs. The above polynomial has the following zeros and multiplicities.<br>• $z_1$ Multiplicity = $i$<br>• $z_2$ Multiplicity = $j$<br>• $z_n$ Multiplicity = $k$ ||
|---|---|---|
| | Odd Multiplicity | Polynomial goes through the zero. |
| | Even Multiplicity | Polynomial bounces at the zero. |
| Example<br><br>Positive "$a$"<br>Even Degree | Bounce — Thru<br>Thru — Bounce ||

## Graphing Polynomial Functions -- Ex. 1

Given: $f(x) = \frac{1}{2}(x+1)^2(x-2)^3(x-4)$

Find: Degree, Zeros ($x$ intercepts), Multiplicity, Far-End behavior, and $y$-intercept. Then sketch it.

| | |
|---|---|
| Degree | $D = 2 + 3 + 1 = 6$  (even) |
| Zeros and Multiplicity | $-1, m2$   $2, m3$   $4, m1$<br>Bounce   Thru.   Thru. |
| Far-End Behavior | Positive leading coefficient and even degree.   ↖ ↗ |
| $y$-intercept | $y$-intercept occurs at $x = 0$<br>$f(0) = \frac{1}{2}(0+1)^2(0-2)^3(0-4)$<br>$f(0) = \frac{1}{2}(1)(-8)(-4) = 16$ |
| Graph | |

## Graphing Polynomial Functions -- Ex. 2

Given: $f(x) = -2x^4(x+3)^2$

Find: Degree, Zeros ($x$ intercepts), Multiplicity, Far-End behavior, and $y$-intercept. Then sketch it.

| | |
|---|---|
| Degree | $D = 4 + 2 = 6$ (even) |
| Zeros and Multiplicity | $-0, m4$     $-3, m2$ <br> Bounce      Bounce |
| Far-End Behavior | Negative leading coefficient and even degree.    ↙ ↘ |
| $y$-intercept | $y$-intercept occurs at $x = 0$ <br> $f(0) = -2(0)^4(0+3)^2$ <br> $f(0) = -2(0)(3) = 0$    Point (0,0) |
| Graph | |

## Graphing Polynomial Functions -- Ex. 3

Given: $f(x) = 2x^2 + 9x - 5$

Find: Degree, Zeros ($x$ intercepts), Multiplicity, Far-End behavior, and $y$-intercept. Then sketch it.

| | |
|---|---|
| Factor it | $f(x) = (2x - 1)(x + 5)$ |
| Degree | $D = 1 + 1 = 2$   (even) |
| Zeros and Multiplicity | $\frac{1}{2}$, $m1$     $-5$, $m1$<br>Thru              Thru |
| Far-End Behavior | Positive leading coefficient and even degree.   ↖ ↗ |
| $y$-intercept | $y$-intercept occurs at $x = 0$<br>$f(0) = (-1)(5) = -5$   Point $(0, -5)$ |
| Graph | |

## Graphing Polynomial Functions -- Ex. 4

Given: $f(x) = 3x^2 - 5x - 2$

Find: Degree, Zeros (x intercepts), Multiplicity, Far-End behavior, and y-intercept. Then sketch it.

| | |
|---|---|
| Factor it | $f(x) = (3x + 1)(x - 2)$ <br> $f(x) = 3\left(x + \frac{1}{3}\right)(x - 2)$ |
| Degree | $D = 1 + 1 = 2$ (even) |
| Zeros and Multiplicity | $-\frac{1}{3}$, m1      2, m1 <br> Thru         Thru |
| Far-End Behavior | Positive leading coefficient and even degree.   ↖ ↗ |
| y-intercept | y-intercept occurs at $x = 0$ <br> $f(0) = -2$     Point: $(0, -2)$ |
| Graph | |

# Graphing Rational Functions

## Graphing Rational Functions

Rational Functions are functions in the form:
$$R(x) = \frac{f(x)}{g(x)} \quad ; g(x) \neq 0$$

$D_n, D_d$ = Degrees of numerator & denom.
$a, b$ = Leading coeff. of numerator & denom.

| Asymptote | # | Look for this | Form |
|---|---|---|---|
| Vertical (VA) | Can be many | Values that make denom. = 0 | $x = c$ |
| Horizontal (HA) | One at most | $D_n < D_d$ | $y = 0$ |
| | | $D_n = D_d$ | $y = \frac{a}{b}$ |
| Slant (SA) | One at most | $D_n = D_d + 1$ | $y = mx + b$ |
| Hole | Can be many | Zeros of factors that cancel. | $(c, 0)$ |

The function will never cross a vertical asymptote.

The function may cross horizontal & slant asymptotes. To check, evaluate: $R(x) = HA$ or $R(x) = SA$

## Graphing Rational Functions – Ex. 1a

Given: $y = \dfrac{2(x-1)(x+2)(x-4)}{3(x-1)(x+3)(x-5)} = \dfrac{2(x+2)(x-4)}{3(x+3)(x-5)}$

Find: All asymptotes, holes, and zeros. Then graph.

| | |
|---|---|
| Degrees | $D_n = 3 \qquad D_d = 3$ |
| Vertical (VA) | $x = -3, 5$ |
| Horizontal (HA) | $y = \dfrac{2}{3} \qquad D_n = D_d$ |
| Slant (SA) | None. |
| Hole | Hole at $x = 1$ $\qquad$ Point: $\left(1, \dfrac{3}{8}\right)$ <br><br> $y = \dfrac{2(1+2)(1-4)}{3(1+3)(1-5)} = \dfrac{-18}{-48} = \dfrac{3}{8}$ |
| $x$ –intercepts | $y = 0 \;\rightarrow\; \text{Numerator} = 0$ <br> $\qquad\quad x = -2, 4$ |
| $y$ –intercept | $x = 0 \;\rightarrow\; y = \dfrac{2(2)(-4)}{3(3)(-5)}$ <br><br> $\qquad\quad y = \dfrac{-16}{-45} \approx 0.35$ |
| Does function cross HA ? | See next page. |

## Graphing Rational Functions – Ex. 1b

Given: $y = \dfrac{2(x-1)(x+2)(x-4)}{3(x-1)(x+3)(x-5)} = \dfrac{2(x+2)(x-4)}{3(x+3)(x-5)}$

Find: All asymptotes, holes, and zeros. Then graph.

| Previously Found | | |
|---|---|---|
| | VA | $x = -3, 5$ |
| | HA | $y = \dfrac{2}{3}$ |
| | Hole | Point: $\left(1, \dfrac{3}{8}\right)$ |
| | $x$ –int'cpt | $y = 0 \to x = -2, 4$ |
| | $y$ –int'cpt | $x = 0 \to y \approx 0.35$ |

| Does function cross HA ? | $\dfrac{2(x+2)(x-4)}{3(x+3)(x-5)} = \dfrac{2}{3}$ |
|---|---|
| | $\dfrac{(x+2)(x-4)}{(x+3)(x-5)} = 1$ |
| | $(x+2)(x-4) = (x+3)(x-5)$ |
| | $x^2 - 2x - 8 = x^2 - 2x - 15$ |
| | $-8 = -15$ |
| | NOT POSSIBLE $\to$ DOES NOT CROSS |

## Graphing Rational Functions – Ex. 1c

Given: $y = \dfrac{2\,(x-1)(x+2)(x-4)}{3\,(x-1)(x+3)(x-5)} = \dfrac{2\,(x+2)(x-4)}{3\,(x+3)(x-5)}$

Find: All asymptotes, holes, and zeros. Then graph.

| | | |
|---|---|---|
| Previously Found | VA | $x = -3,\ 5$ |
| | HA | $y = \dfrac{2}{3}$   Does NOT cross |
| | Hole | Point: $\left(1,\ \dfrac{3}{8}\right)$ |
| | $x$ –int'cpt | $y = 0 \ \rightarrow\ x = -2,\ 4$ |
| | $y$ –int'cpt | $x = 0 \ \rightarrow\ y \approx 0.35$ |

| | |
|---|---|
| Graph | 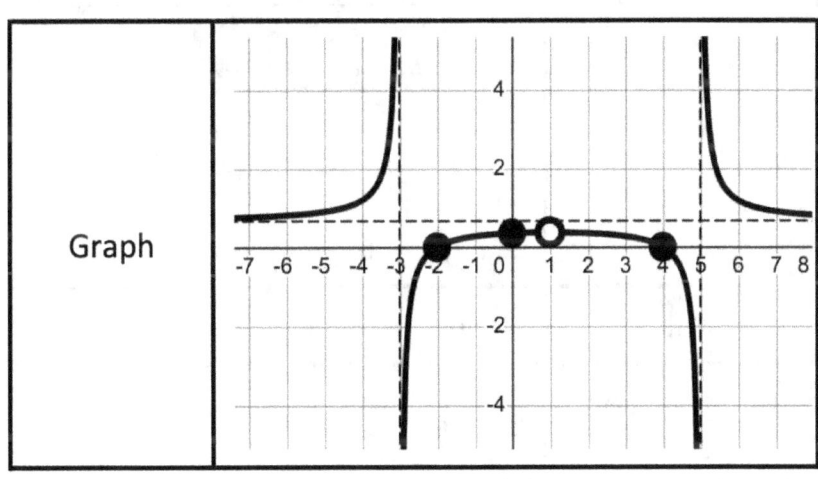 |

| | **Graphing Rational Functions – Ex. 2a** |
|---|---|
| Given: $y = \dfrac{6x^2 + 7}{x + 1}$ <br> Find: All asymptotes, holes, and zeros. Then graph. ||
| Degrees | $D_n = 2 \quad D_d = 1$ |
| Vertical (VA) | $x = -1$ |
| Horizontal (HA) | None. |
| Slant (SA) | $D_n = D_d + 1 \rightarrow$ There is a SA <br><br> Divide numerator by denominator and ignore the remainder. <br><br> $\begin{array}{r} 6x - 6 \phantom{00} \\ x+1 \overline{\smash{)}\, 6x^2 + 0x + 7} \\ -\underline{(6x^2 - 6x)\phantom{00}} \\ 6x + 7 \\ -\underline{(-6x - 6)} \\ 13 \end{array}$ <br><br> SA: $y = 6x - 6$ |
| Hole | None. |
| Does function cross SA ? | See next page. |

| Graphing Rational Functions – Ex. 2b | | |
|---|---|---|
| Given: $y = \dfrac{6x^2 + 7}{x + 1}$<br>Find: All asymptotes, holes, and zeros. Then graph. | | |
| Previously Found | VA | $x = -1$ |
|  | SA | $y = 6x - 6$ |
| Does function cross SA ? | \multicolumn{2}{l}{$\dfrac{6x^2 + 7}{x + 1} = 6x - 6$ <br><br> $6x^2 + 7 = (6x - 6)(x + 1)$ <br> $6x^2 + 7 = 6(x - 1)(x + 1)$ <br> $6x^2 + 7 = 6(x^2 - 1)$ <br> $6x^2 + 7 = 6x^2 - 6$ <br> $7 = 6$ <br> NOT POSSIBLE $\rightarrow$ Does not cross SA} |
| $x$ –intercepts | \multicolumn{2}{l}{$y = 0 \rightarrow$ Numerator $= 0$ <br> $6x^2 + 7 = 0$ <br> $x = \pm\sqrt{-\dfrac{7}{6}}$ \quad Imaginary number <br> No x-intercepts} |
| $y$ –intercept | \multicolumn{2}{l}{$x = 0 \rightarrow y = \dfrac{0 + 7}{0 + 1} = 7$ <br> Point: $(0, 7)$} |

### Graphing Rational Functions – Ex. 2c

Given: $y = \dfrac{6x^2 + 7}{x + 1}$

Find: All asymptotes, holes, and zeros. Then graph.

| Previously Found | | |
|---|---|---|
| | VA | $x = -1$ |
| | SA | $y = 6x - 6$   Does not cross |
| | $x$ Int'cpt | No x intercepts |
| | $y$ Int'cpt | Point: $(0, 7)$ |

Graph

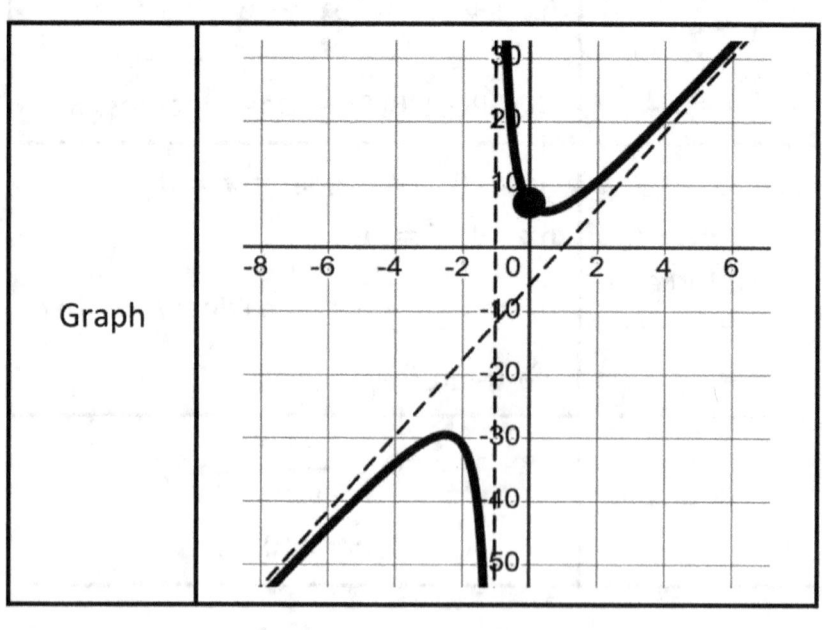

## Graphing Rational Functions – Ex. 3a

Given: $y = \dfrac{1}{x}$

Find: All asymptotes, holes, and zeros. Then graph.

| | |
|---|---|
| Degrees | $D_n = 0 \qquad D_d = 1$ |
| Vertical (VA) | $x = 0 \qquad$ Makes denom. $= 0$ |
| Horizontal (HA) | $y = 0 \qquad D_d > D_n$ |
| Slant (SA) | None. |
| Holes | None. |
| Does function cross HA ? | $\dfrac{1}{x} = 0$ <br> $1 = 0 \qquad$ NOT Possible <br> Function does NOT cross HA |
| x-intercepts | $y = 0 \;\to\;$ Numerator $= 0$ <br> NOT Possible. $\to$ No x-intercepts |
| y-intercept | $x = 0 \;\to\; y = \dfrac{1}{x} = \dfrac{1}{0}$ <br> NOT Possible $\to$ No y-intercept |
| Graph | Next page… |

## Graphing Rational Functions – Ex. 3b

Given: $y = \dfrac{1}{x}$

Find: All asymptotes, holes, and zeros. Then graph.

| Previously Found | VA | $x = 0$ | Multiplicity = 1  Odd Multiplicity  Function up on one side and down on other. |
| --- | --- | --- | --- |
| | HA | $y = 0$ | Does not cross. |
| | x-int'cpt | None. | |
| | y-int'cpt | None. | |

| Graph | 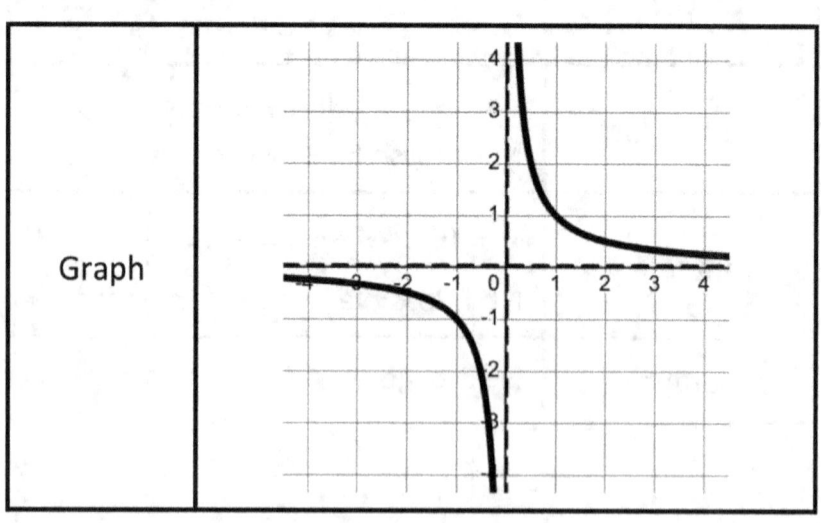 |
| --- | --- |

## Graphing Rational Functions – Ex. 4a

Given: $y = \dfrac{1}{x^2}$

Find: All asymptotes, holes, and zeros. Then graph.

| | |
|---|---|
| Degrees | $D_n = 0$    $D_d = 2$ |
| Vertical (VA) | $x = 0$    Makes denom. $= 0$ |
| Horizontal (HA) | $y = 0$    $D_d > D_n$ |
| Slant (SA) | None. |
| Holes | None. |
| Does function cross HA ? | $\dfrac{1}{x^2} = 0$<br>$1 = 0$    NOT Possible<br>Function does NOT cross HA |
| x-intercepts | $y = 0$ → Numerator $= 0$<br>NOT Possible. → No x-intercepts |
| y-intercept | $x = 0$ → $y = \dfrac{1}{x} = \dfrac{1}{0}$<br>NOT Possible → No y-intercept |
| Graph | Next page... |

## Graphing Rational Functions – Ex. 4b

Given: $y = \dfrac{1}{x^2}$

Find: All asymptotes, holes, and zeros. Then graph.

| Previously Found | VA | $x = 0$ | Multiplicity = 2  Even Multiplicity. Function goes both up or both down near VA. |
|---|---|---|---|
| | HA | $y = 0$ | Does not cross. |
| | x-int'cpt | None. | |
| | y-int'cpt | None. | |

Graph

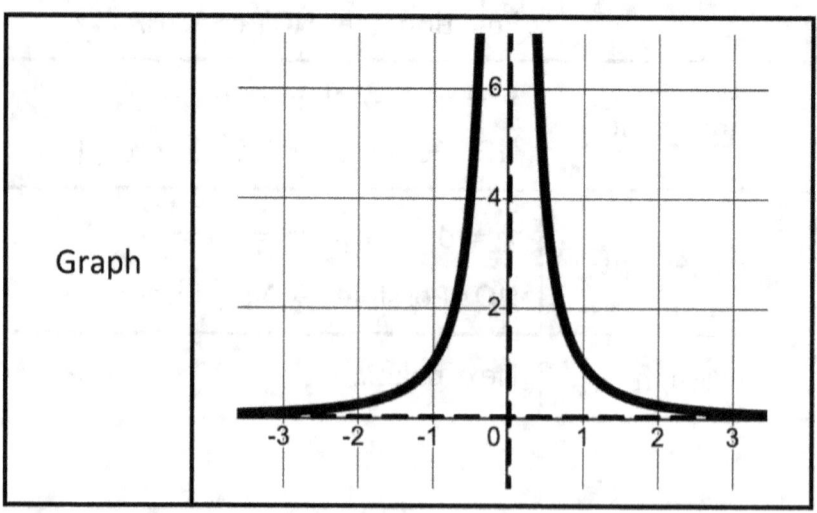

## Graphing Rational Functions – Ex. 5a

Given: $f(x) = \dfrac{x^2 - 4}{x^2 - 4x} = \dfrac{(x-2)(x+2)}{x(x-4)}$

Find: All asymptotes, holes, and zeros. Then graph.

| | |
|---|---|
| Degrees | $D_n = 2 \quad D_d = 2$ |
| Vertical (VA) | $x = 0$ and $x = 4$ |
| Horizontal (HA) | $y = \dfrac{1}{1} = 1$ |
| Slant (SA) | None. |
| Hole | None. |
| Does function cross HA ? | $\dfrac{x^2 - 4}{x^2 - 4x} = 1$ <br><br> $x^2 - 4 = x^2 - 4x$ <br><br> $-4 = -4x$ <br><br> $\dfrac{-4}{-4} = \dfrac{-4x}{-4}$ <br><br> $1 = x \qquad$ This is possible. <br><br> Function crosses HA at $x = 1$ <br><br> $f(1) = \dfrac{(1-2)(1+2)}{1(1-4)} = \dfrac{-3}{-3} = 1$ <br><br> Point: $(1, 1)$ |

## Graphing Rational Functions – Ex. 5b

Given: $f(x) = \dfrac{x^2 - 4}{x^2 - 4x} = \dfrac{(x-2)(x+2)}{x(x-4)}$

Find: All asymptotes, holes, and zeros. Then graph.

| Previously Found | VA | $x = 0$, $x = 4$ |
| --- | --- | --- |
| | HA | $y = 1$  Function crosses at point: $(1, 1)$ |
| x-intercepts | | $y = 0 \rightarrow$ Numerator $= 0$ <br> $x = \pm 2$ |
| y-intercept | | $x = 0 \rightarrow y = \dfrac{0^2 - 4}{0^2 - 4(0)} = \dfrac{-4}{0}$ DNE <br> No y-intercept |
| Graph | | 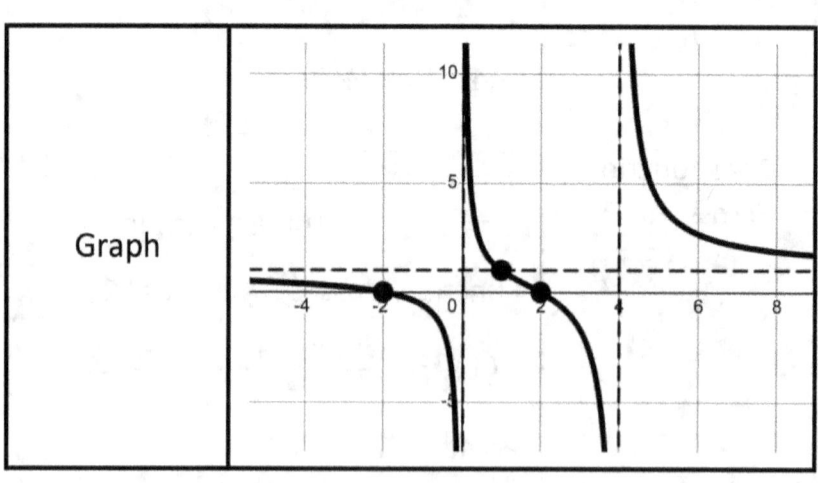 |

| Graphing Rational Functions – Ex. 6a ||
|---|---|
| Given: $f(x) = \dfrac{3x^2 - 4x}{x^2 + 6}$ <br> Find: All asymptotes, holes, and zeros. Then graph. ||
| Degrees | $D_n = 2 \quad D_d = 2$ |
| Vertical (VA) | None. |
| Horizontal (HA) | $y = \dfrac{3}{1} = 3$ |
| Slant (SA) | None. |
| Hole | None. |
| Does function cross HA ? | $\dfrac{3x^2 - 4x}{x^2 + 6} = 3$ <br><br> $3x^2 - 4x = 3x^2 + 18$ <br><br> $-4x = 18$ <br><br> $x = \dfrac{18}{-4} = -\dfrac{9}{2} = -4.5$ <br><br> $x = -4.5 \quad$ This is possible. <br><br> Function crosses HA at $x = -4.5$ <br><br> Point: $(-4.5, 3)$ |

## Graphing Rational Functions – Ex. 6b

Given: $f(x) = \dfrac{3x^2 - 4x}{x^2 + 6}$

Find: All asymptotes, holes, and zeros. Then graph.

| | | |
|---|---|---|
| Previously Found | VA | None. |
| | HA | $y = 3$<br>Function crosses HA at Point: $(-4.5, 3)$ |
| x-intercepts | | $y = 0 \rightarrow$ Numerator $= 0$<br>$x(3x - 4) = 0$<br>$x = 0, \dfrac{4}{3}$<br>Points: $(0,0)$, $\left(\dfrac{4}{3}, 0\right)$ |
| y-intercept | | $x = 0 \rightarrow y = \dfrac{0}{6} = 0$    Point: $(0, 0)$ |

| | |
|---|---|
| Graph | 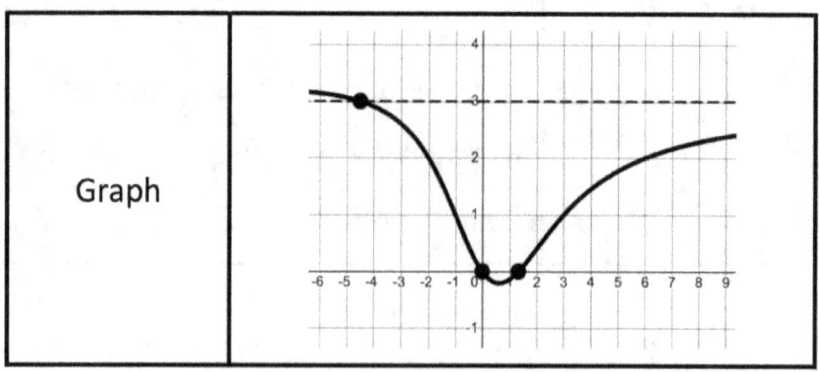 |

# Working With Radicals

## Rationalize the Denominator

Radicals in the denominator are often removed by **Rationalizing the Denominator**. Radicals in the numerator are okay.

Recall: $a^2 - b^2 = (a-b)(a+b)$

And: $\left[\sqrt{a}\right]^2 = a$

| | | |
|---|---|---|
| $y = \dfrac{1}{\sqrt{123}}$ | $y = \dfrac{1}{\sqrt{123}}$ (1) | |
| | $y = \dfrac{1}{\sqrt{123}}\left(\dfrac{\sqrt{123}}{\sqrt{123}}\right) = \dfrac{\sqrt{123}}{123}$ | |
| $y = \dfrac{1}{2+\sqrt{3}}$ | $y = \dfrac{1}{2+\sqrt{3}}$ (1) | |
| | $y = \dfrac{1}{2+\sqrt{3}}\left(\dfrac{2-\sqrt{3}}{2-\sqrt{3}}\right)$ | |
| | $y = \dfrac{2-\sqrt{3}}{4-3} = \dfrac{2-\sqrt{3}}{1} = 2-\sqrt{3}$ | |
| $y = \dfrac{1}{6-\sqrt{7}}$ | $y = \dfrac{1}{6-\sqrt{7}}$ (1) | |
| | $y = \dfrac{1}{6-\sqrt{7}}\left(\dfrac{6+\sqrt{7}}{6+\sqrt{7}}\right)$ | |
| | $y = \dfrac{6+\sqrt{7}}{36-7} = \dfrac{6+\sqrt{7}}{29}$ | |

## Working With Radicals

Recall, a radical is in the form: $\sqrt[n]{Radicand}$
Where: $n$ = index
Default: $n = 2$    Therefore: $\sqrt[2]{25} = \sqrt{25} = 5$

| Even $n$ | $y = \sqrt[n]{x}$ | 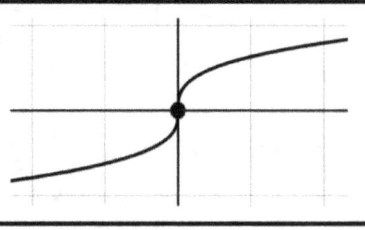 |
|---|---|---|
| | $D: [\,0, \infty\,)$ $R: [\,0, \infty\,)$ | |
| Odd $n$ | $y = \sqrt[n]{x}$ | |
| | $D: (-\infty, \infty\,)$ $R: (-\infty, \infty\,)$ | |

| | Notes: |
|---|---|
| When $n$ is even: The radicand $\geq 0$ | $\sqrt{9} = 3$ $\sqrt{-9} = Undefined$ $\sqrt[3]{8} = 2$ $\sqrt[3]{-8} = -2$ |
| When solving equations with radicals, isolate one radical and solve. | Note: $[\,\sqrt[n]{f(x)}\,]^n = f(x)$ |

## Working With Radicals -- Ex. 1

Given: $f(x) = \sqrt{3x + 12}$
Find: Domain and Range. Then, graph it.
Note: Using translations may help.

| | | |
|---|---|---|
| Domain & Range | $n =$ even<br>Radicand $\geq 0$<br>$3x + 12 \geq 0$<br>$x \geq -\frac{12}{3}$<br>$D: [-4, \infty)$ | $f(x) = \sqrt{3x + 12}$<br>$f(-4) = 0$<br>$R: [0, \infty)$ |
| Translations<br><br>Parent:<br>$y = \sqrt{x}$ | $y = \sqrt{3x + 12}$<br>$y = \sqrt{3(x + 4)}$ | H. Comp. by $\frac{1}{3}$<br>H. Shift left by 4<br>No Vert. Trans. |
| Graph | | |

| Equations With Radicals -- Ex. 1 |
|---|
| Solve for $x$: $\quad 3x - 2 = 5\sqrt{x}$ <br> Check your answer(s) for extraneous solutions. |

| | | |
|---|---|---|
| Isolate term with radical <br><br> Then square both sides | $3x - 2 = 5\sqrt{x}$ <br> $[3x - 2]^2 = [5\sqrt{x}]^2$ <br> $(3x - 2)(3x - 2) = 25x$ <br> $9x^2 - 12x + 4 = 25x$ <br> $9x^2 - 37x + 4 = 0$ <br> $x = \dfrac{37 \pm \sqrt{(-37)^2 - 4(9)(4)}}{2(9)}$ <br> $x = \dfrac{37 \pm \sqrt{1225}}{18} = \dfrac{37 \pm 35}{18}$ <br> $x = \dfrac{72}{18}, \dfrac{2}{18} = 4, \dfrac{1}{9}$ | |
| Check answers <br><br> Use Original Equation | $x = 4$ | $3(4) - 2 = 5\sqrt{4}$ <br> $10 = 10 \qquad$ TRUE |
| | $x = \dfrac{1}{9}$ | $3\left(\dfrac{1}{9}\right) - 2 = 5\sqrt{\dfrac{1}{9}}$ <br> $-\dfrac{5}{3} = \dfrac{5}{3} \qquad$ FALSE |
| Solution | $x = 4$ | |

## Equations With Radicals -- Ex. 3a

Solve for $x$: $\sqrt{2x+5} - 10 = 2\sqrt{2x} - 9$
Check your answer(s) for extraneous solutions.

Recall: $(a+b)^2 = a^2 + 2ab + b^2$
$(a-b)^2 = a^2 - 2ab + b^2$

Isolate one radical term then square both sides.

Do this twice.

$$\sqrt{2x+5} = 2\sqrt{2x} + 1$$
$$\left[\sqrt{2x+5}\right]^2 = \left[2\sqrt{2x}+1\right]^2$$
$$2x + 5 = (2\sqrt{2x}+1)(2\sqrt{2x}+1)$$
$$2x + 5 = 4(2x) + 4\sqrt{2x} + 1$$
$$2x + 5 = 8x + 4\sqrt{2x} + 1$$
$$-6x + 4 = 4\sqrt{2x}$$
$$-3x + 2 = 2\sqrt{2x}$$
$$[2-3x]^2 = \left[2\sqrt{2x}\right]^2$$
$$4 - 12x + 9x^2 = 4(2x)$$
$$9x^2 - 20x + 4 = 0$$

Quadratic Equation

$$x = \frac{20 \pm \sqrt{(-20)^2 - 4(9)(4)}}{2(9)}$$

$$x = \frac{20 \pm \sqrt{256}}{18} = \frac{20 \pm 16}{18} = 2, \frac{2}{9}$$

## Equations With Radicals -- Ex. 3b

**Solve for $x$:** $\sqrt{2x+5} - 10 = 2\sqrt{2x} - 9$
Check your answer(s) for extraneous solutions.

Recall: $(a+b)^2 = a^2 + 2ab + b^2$
$(a-b)^2 = a^2 - 2ab + b^2$

| Previously Found | $x = 2, \frac{2}{9}$ | |
|---|---|---|
| Check answers<br><br>Use Original Equation | $x = 2$ | $\sqrt{2x+5} = 2\sqrt{2x} + 1$<br>$\sqrt{4+5} = 2\sqrt{4} + 1$<br>$3 = 4 + 1$<br>$3 = 5$     FALSE |
| | $x = \frac{2}{9}$ | $\sqrt{2x+5} = 2\sqrt{2x} + 1$<br>$\sqrt{\frac{4}{9} + \frac{45}{9}} = 2\sqrt{\frac{4}{9}} + 1$<br>$\sqrt{\frac{49}{9}} = 2\left(\frac{2}{3}\right) + 1$<br>$\frac{7}{3} = \frac{4}{3} + \frac{3}{3}$     TRUE |
| Solution | $x = \frac{2}{9}$ | |

## Equations With Radicals -- Ex. 4a

| | | |
|---|---|---|
| Solve for $x$: | $\dfrac{2x}{\sqrt{x-1}} = x$ | |
| Domain | $x - 1 > 0$ <br> $x > 1$ | Radicand must be $\geq 0$ <br> AND Denominator $\neq 0$ <br> So, $x > 1$ |
| Here, we can start by squaring both sides. | $\left[\dfrac{2x}{\sqrt{x-1}}\right]^2 = [x]^2$ <br><br> $\dfrac{4x^2}{x-1} = x^2$ <br><br> $4x^2 = x^2(x-1)$ | |
| | You can not divide by $x$ because $x$ might $= 0$. Division by zero is undefined. | |
| Don't divide by $x$. Factor instead. | $4x^2 = x^3 - x^2$ <br><br> $0 = x^3 - 5x^2$ <br><br> $x^2(x-5) = 0$ <br><br> $x = 0, 5$ | |
| Note | $x = 0$ is not in the domain. <br> If $x = 0$, denominator is imaginary. <br> Therefore, $x = 5$ | |

| Equations With Radicals -- Ex. 4b ||
|---|---|
| Solve for $x$: $\dfrac{2x}{\sqrt{x-1}} = x$ ||

| | Domain | $x > 1$ |
|---|---|---|
| Previously Found | | |
| | One Solution in domain | $x = 5$ |
| Check | $x = 5 \rightarrow$ $\dfrac{2x}{\sqrt{x-1}} = x$ $\dfrac{2(5)}{\sqrt{5-1}} = 5$ $\dfrac{10}{\sqrt{4}} = 5$ $\dfrac{10}{2} = 5$ | TRUE |

| **Equations With Radicals -- Ex. 5** ||
|---|---|
| Solve for $x$: $\sqrt[3]{2x-5} + 7 = 4$ ||
| Isolate one radical term then cube both sides. | $\sqrt[3]{2x-5} = -3$ <br><br> $\left[\sqrt[3]{2x-5}\right]^3 = [-3]^3$ <br><br> $2x - 5 = (-3)(-3)(-3)$ <br><br> $2x - 5 = -27$ <br><br> $2x = -22$ <br><br> $x = -11$ |
| No need to check for extraneous roots with cube roots (odd index) <br><br> But check anyway! | $\sqrt[3]{2x-5} + 7 = 4$ <br><br> $\sqrt[3]{2(-11)-5} + 7 = 4$ <br><br> $\sqrt[3]{-22-5} + 7 = 4$ <br><br> $\sqrt[3]{-27} + 7 = 4$ <br><br> $(-3) + 7 = 4$ <br><br> $4 = 4$        TRUE |

## The Inverse of $y = \sqrt{x}$ -- Ex. 6

| | | |
|---|---|---|
| Given: $f(x) = \sqrt{x}$ <br> Find: Domain, range, and inverse of $f(x)$ <br> And: Domain and range of $f^{-1}(x)$ <br> Then: Sketch both $f(x)$ & $f^{-1}(x)$ on same graph. | | |
| $y = f(x)$ | $y = \sqrt{x}$ | D: $[0, \infty)$  R: $[0, \infty)$ |
| Find the Inverse of $f(x)$ | $y = \sqrt{x}$ <br> $x = \sqrt{y}$ <br> $x^2 = y$  $\rightarrow$ | Original Equation <br> Switch $x$ and $y$ <br> $f^{-1}(x) = x^2$ |
| Let $g = f^{-1}(x)$ | $g = x^2$ <br> Domain & Inverses are switched. <br> They're the same, so not a big deal. | D: $[0, \infty)$  R: $[0, \infty)$ |
| Graph <br> $y = f(x)$ <br> $g = f^{-1}(x)$ | 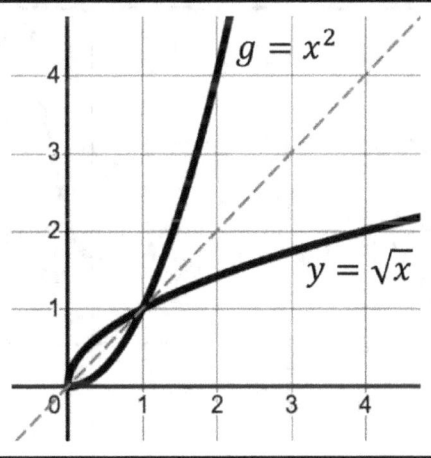 | |

| The Inverse of $y = \sqrt{x-5}$ -- Ex. 7 ||
|---|---|
| Given: $f(x) = \sqrt{x-5}$ <br> Find: Domain, range, and inverse of $f(x)$ <br> And: Domain and range of $f^{-1}(x)$ <br> Then: Sketch both $f(x)$ & $f^{-1}(x)$ on same graph. ||
| $y = f(x)$ | $y = \sqrt{x-5}$    D: $[5, \infty)$    R: $[5, \infty)$ |
| Find the Inverse of $f(x)$ | $y = \sqrt{x-5}$    Original Equation <br> $x = \sqrt{y-5}$    Switch $x$ and $y$ <br> $x^2 = y - 5$    $\rightarrow$    $f^{-1}(x) = x^2 + 5$ |
| Let $g = f^{-1}(x)$ | $g = x^2 + 5$    D: $[5, \infty)$    R: $[5, \infty)$ <br> Domain & Inverses are switched. <br> They're the same, so not a big deal. |
| Graph <br> $y = f(x)$ <br> $g = f^{-1}(x)$ | 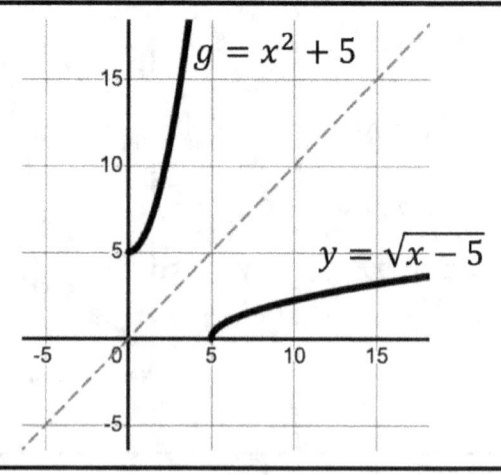 |

## The Inverse of $y = \sqrt[3]{x-5}$ -- Ex. 8

**Given:** $f(x) = \sqrt[3]{x-5}$
**Find:** Domain, range, and inverse of $f(x)$
**And:** Domain and range of $f^{-1}(x)$
**Then:** Sketch both $f(x)$ & $f^{-1}(x)$ on same graph.

| | | |
|---|---|---|
| $y = f(x)$ | D: $(-\infty, \infty)$ | R: $(-\infty, \infty)$ |
| Find the Inverse of $f(x)$ | $y = \sqrt[3]{x-5}$ <br> $x = \sqrt[3]{y-5}$ <br> $x^3 = y - 5 \;\rightarrow\;$ | Original Equation <br> Switch $x$ and $y$ <br> $f^{-1}(x) = x^3 + 5$ |
| Let <br> $g = f^{-1}(x)$ <br> $g = x^3 + 5$ | D: $(-\infty, \infty)$    R: $(-\infty, \infty)$ <br> Domain & Inverses are switched. <br> They're the same, so not a big deal. | |
| Graph <br><br> $y = f(x)$ <br> $g = f^{-1}(x)$ | $g = x^3 + 5$ <br> $y = \sqrt[3]{x-5}$ | |

# Logarithms

| | Logarithm Properties |
|---|---|
| Definition of a Log | $log_b(a) = n \iff b^n = a$ |
| Reference Points $(1,0)$ & $(b,1)$ | $log_b(a) = n \iff b^n = a$ <br> $log_b(b) = 1 \iff b^1 = b$ |
| Conventions | $log_{10}(a) = \log(a)$ <br> $log_e(a) = \ln(a)$ |
| Product Rule | $\log(a) + \log(b) = \log(ab)$ |
| Quotient Rule | $\log(a) - \log(b) = \log\left(\frac{a}{b}\right)$ |
| Power Rule | $\log(a^n) = n \cdot \log(a)$ |
| Reciprocal Rule | $\log\left(\frac{1}{a}\right) = \log(a^{-1}) = -\log(a)$ |
| Inverse Properties | $log_b(b^x) = x$ <br> $b^{log_b(x)} = x$ |
| Change of Base | $log_b(a) = \frac{log_c(a)}{log_c(b)} = \frac{\log(a)}{\log(b)}$ |

## Logarithm Graphs

| | |
|---|---|
| $y = \log_b(x)$<br>$b > 1$ | 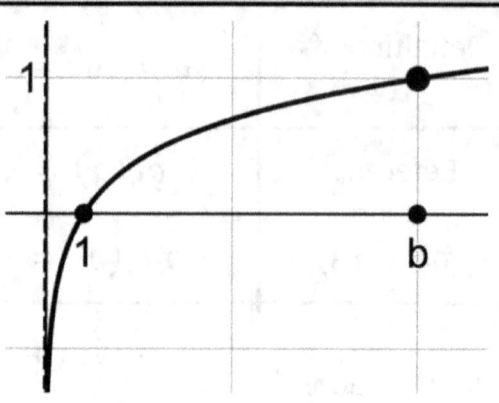 |
| $y = \log_b(x)$<br>$b < 1$ | 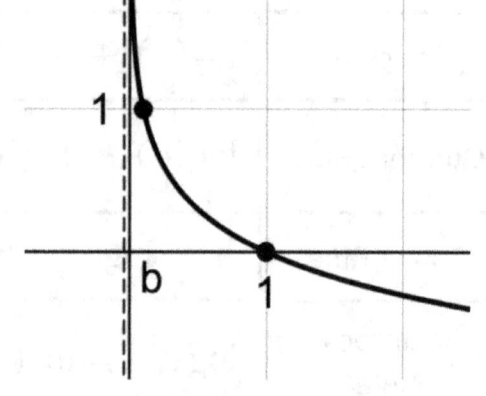 |

Domain: $(0, \infty)$

Range: $(-\infty, \infty)$

Asymptote: $x = 0$

$\log_b(1) = 0$    $\log_b(b) = 1$

## Logarithms -- Ex. 1

Use the definition of a log to convert the following logs to exponential form.

$$log_b(a) = n \iff b^n = a$$

| Log | Exponential Form |
|---|---|
| $log_9(81) = 2$ | $9^2 = 81$ |
| $\ln 1 = 0$ | $e^0 = 1$ |
| $\log 1 = 0$ | $10^0 = 1$ |
| $log_7(1) = 0$ | $7^0 = 1$ |
| $log_b(12) = 34$ | $b^{34} = 12$ |
| $log_2(8) = 3$ | $2^3 = 8$ |
| $log_q(r) = s$ | $q^s = r$ |
| $\log 10 = 1$ | $10^1 = 10$ |
| $\ln e = 1$ | $e^1 = e$ |
| $log_b(b) = 1$ | $b^1 = b$ |

## Logarithms -- Ex. 2

Use the definition of a log to convert the following from exponential form to log form.

| Exponential Form | Log Form |
|---|---|
| $5^3 = 125$ | $log_5(125) = 3$ |
| $\left(\frac{1}{5}\right)^{-3} = 125$ | $log_{\frac{1}{5}}(125) = -3$ |
| $10^5 = 100{,}000$ | $\log(100{,}000) = 5$ |
| $b^0 = 1$ | $log_b(1) = 0$ |
| $b^9 = n$ | $log_b(n) = 9$ |
| $2^3 = 8$ | $log_2(8) = 3$ |
| $10^1 = 10$ | $\log 10 = 1$ |
| $e^1 = e$ | $\ln e = 1$ |
| $b^1 = b$ | $log_b(b) = 1$ |
| $12^n = 34$ | $log_{12}(34) = 12$ |
| $10^n = 123$ | $\log 123 = n$ |

## Logarithms -- Ex. 3

Simplify the expressions without using a calculator.

Recall: $\log(a^n) = n \cdot \log(a)$ and $\log_b(b) = 1$

| Expression | Simplified |
|---|---|
| $\log_3(9)$ | $\log_3(3^2)$ <br> $2 \cdot \log_3(3)$ <br> $2 \cdot (1) \;=\; 2$ |
| $\log_2(16)$ | $\log_2(2^4)$ <br> $4 \cdot \log_2(2)$ <br> $4 \cdot (1) \;=\; 4$ |
| $\log_2\left(\frac{1}{16}\right)$ | $\log_2(2^{-4})$ <br> $-4 \cdot \log_2(2)$ <br> $-4 \cdot (1) \;=\; -4$ |
| $\log_{\frac{1}{6}}\left(\frac{1}{36}\right)$ | $\log_{\frac{1}{6}}\left(\left(\frac{1}{6}\right)^2\right)$ <br> $2 \cdot \log_{\frac{1}{6}}\left(\frac{1}{6}\right)$ <br> $2 \cdot (1) \;=\; 2$ |

## Graphs of Log Functions -- Ex. 4a

For each given function: Graph it and identify the VA. Also state the domain and range in interval notation.
<u>Hint:</u> Plot points $(1, 0)$ & $(b, 1)$ to help with graphing.

---

$y = \log_3(x + 2)$

H. Shift left by 2

No V. Translations

| $x - 2$ | $x$ | $y$ | $y$ |
|---|---|---|---|
| $-1$ | 1 | 0 | 0 |
| 1 | 3 | 1 | 1 |

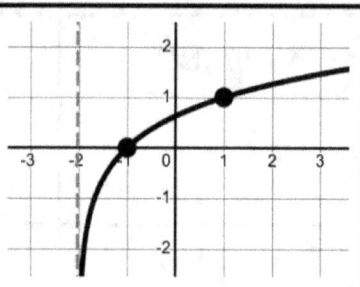

D: $(-2, \infty)$
R: $(-\infty, \infty)$
VA: $x = -2$

---

$y = \log_3(x - 1) - 2$

H. Shift right by 1

V. Shift down by 2

| $x + 1$ | $x$ | $y$ | $y - 2$ |
|---|---|---|---|
| 2 | 1 | 0 | $-2$ |
| 4 | 3 | 1 | $-1$ |

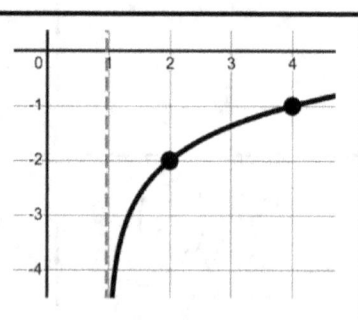

D: $(1, \infty)$
R: $(-\infty, \infty)$
VA: $x = 1$

## Graphs of Log Functions -- Ex. 4b

For each given function: Graph it and identify the VA. Also state the domain and range in interval notation.
<u>Hint:</u> Plot points $(1, 0)$ & $(b, 1)$ to help with graphing.

### $y = -\log_4 x$

V. Rotation over x-axis

No H. Translations

| $x$ | $x$ | $y$ | $-y$ |
|---|---|---|---|
| 1 | 1 | 0 | 0 |
| 4 | 4 | 1 | $-1$ |

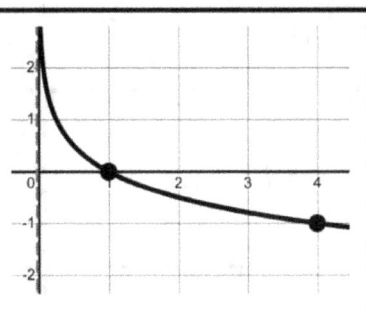

D: $(0, \infty)$
R: $(-\infty, \infty)$
VA: $x = 0$

### $y = \log_2(x - 1)$

H. Shift right by 1

No V. Translations

| $x+1$ | $x$ | $y$ | $y$ |
|---|---|---|---|
| 2 | 1 | 0 | 0 |
| 3 | 2 | 1 | 1 |

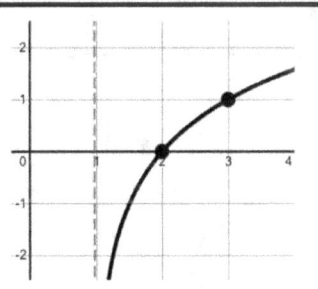

D: $(1, \infty)$
R: $(-\infty, \infty)$
VA: $x = 1$

## Domains of Log Functions -- Ex. 5a

For each function write the domain in interval notation.

| Function | Calculations | Domain |
|---|---|---|
| $y = \log_3(x - 7)$ | $x - 7 > 0$ <br> $x > 7$ | $(7, \infty)$ |
| $y = \log_2(x^2 + 5)$ | $x^2 + 5 > 0$ <br> $x^2 > -5$ <br> True for all $x$ | $(-\infty, \infty)$ |
| $y = \log_7(x^2 - 9)$ | $x^2 - 9 > 0$ <br> $x^2 > 9$ <br> $x < -3$ or $x > 3$ | $(-\infty, -3)$ <br> U <br> $(3, \infty)$ |
| $y = \log_{\frac{1}{6}}(x + 2)$ | $x + 2 > 0$ <br> $x > -2$ | $(-2, \infty)$ |
| $y = \ln(x^2)$ | $x^2 > 0$ <br> $x \neq 0$ | $(-\infty, 0)$ <br> U <br> $(0, \infty)$ |
| $y = \log(x + 2) + 1$ | $x + 2 > 0$ <br> $x > -2$ | $(-2, \infty)$ |

## Domains of Functions -- Ex. 5b

For each function write the domain in interval notation.

| Function | Calculations | Domain |
|---|---|---|
| $y = \log(x)$ | $x > 0$ | $(0, \infty)$ |
| $y = \sqrt{x}$ | $x \geq 0$ | $[0, \infty)$ |
| $y = \dfrac{1}{x}$ | $x \neq 0$ | $(-\infty, 0)$ ∪ $(0, \infty)$ |
| $y = \dfrac{1}{\sqrt{x}}$ | $x \geq 0$ AND $x \neq 0$ Therefore: $x > 0$ | $(0, \infty)$ |

## Log Sums and Differences -- Ex. 6

Write the logarithm as a sum or difference of logs. Simplify as much as possible.

| Logarithm | Written as sums & differences. |
|---|---|
| $\log_a a^2$ | $2 \log_a a$ <br> $2(1) = 2$ |
| $\log_4(AB)$ | $\log_4 A + \log_4 B$ |
| $\log_4\left(\frac{1}{16} x^3 z\right)$ | $\log_4\left(\frac{1}{16}\right) + \log_4 x^3 + \log_4 z$ <br> $\log_4(4^{-2}) + \log_4 x^3 + \log_4 z$ <br> $-2\log_4(4) + 3\log_4 x + \log_4 z$ <br> $-2 + 3\log_4 x + \log_4 z$ |
| $\log\left(\frac{100}{\sqrt{a^2+b^2}}\right)$ | $\log(100) - \log(\sqrt{a^2+b^2})$ <br> $\log(10^2) - \log(a^2+b^2)^{\frac{1}{2}}$ <br> $2\log(10) - \frac{1}{2}\log(a^2+b^2)$ <br> $2 - \frac{1}{2}\log(a^2+b^2)$ |
| $\log\left(\frac{AB}{CD}\right)$ | $\log A + \log B - \log C - \log D$ |

## Log Sums and Differences -- Ex. 7

Write the logarithm expression as a single log with a coefficient of 1. Simplify as much as possible.

| Logarithm | Written as a single log. |
|---|---|
| $\ln x + \ln 4$ | $\ln(4x)$ |
| $\log_6 144 - \log_6 4$ | $\log_6\left(\frac{144}{4}\right)$ <br> $\log_6(36)$ <br> $\log_6(6^2)$ <br> $2 \cdot \log_6(6) \;=\; 2$ |
| $5\log_3 x + \log_3 z$ | $\log_3(x^5) + \log_3(z)$ <br> $\log_3(x^5 z)$ |
| $\frac{1}{2}\ln(x^2 - 1) - \frac{1}{2}\ln(x + 1)$ | $\frac{1}{2}\left[\ln\left(\frac{x^2-1}{x+1}\right)\right]$ <br> $\frac{1}{2}\left[\ln\left(\frac{(x-1)(x+1)}{x+1}\right)\right]$ <br> $\frac{1}{2}[\ln(x-1)]$ <br> $\ln(x-1)^{\frac{1}{2}}$ <br> $\ln\sqrt{x-1}$ |

## Exponential Equations -- Ex. 8

Solve the exponential equations.

| Exponential Equation | Solution |
|---|---|
| $3^x = 81$ | $3^x = 81$ <br> $3^x = 3^4$ <br> $x = 4$ |
| $5^{2x+3} = 625$ | $5^{2x+3} = 625$ <br> $5^{2x+3} = 5^4$ <br> $2x + 3 = 4$ <br> $x = \frac{1}{2}$ |
| $8^{10x} = 5^{100}$ | $8^{10x} = 5^{100}$ <br> $\log(8^{10x}) = \log(5^{100})$ <br> $(10x)\log(8) = (100)\log(5)$ <br> $\frac{(10x)}{(100)} = \frac{\log(5)}{\log(8)} = \log_8 5 \approx .774$ <br> $\frac{x}{10} \approx .774$ <br> $x \approx 7.74$ |
| $4^x = 24$ | $\log_4(24) = x$    Definition of a log. |

## Solving Log Equations -- Ex. 9a

Solve the logarithmic equations.

---

$log_4(x + 1) = 3$

| | |
|---|---|
| $4^3 = x + 1$ | Definition of a Log |
| $x + 1 = 64$ | |
| $x = 63$ | |

---

$\log(3x + 11) = \log(3 - x)$

$3x + 11 = 3 - x$

$4x = -7$

$x = -\dfrac{7}{4}$

---

$log_4(x^2 + 30x) = 3$

| | |
|---|---|
| $4^3 = x^2 + 30x$ | Definition of a Log |
| $x^2 + 30x - 64 = 0$ | |
| $(x - 2)(x + 32) = 0$ | |
| $x = 2, -32$ | |

**Solving Log Equations -- Ex. 9b**

Solve the logarithmic equations.

---

$5 \log_3(7 - 5x) + 2 = 17$

$5 \log_3(7 - 5x) = 15$

$\log_3(7 - 5x) = 3$

$3^3 = (7 - 5x)$

$5x = 7 - 3^3$

$x = \dfrac{7-27}{5} = \dfrac{-20}{5} = -4$

---

$\log x + \log(x - 10) = \log(x - 18)$

$\log(x(x - 10)) = \log(x - 18)$

$x^2 - 10x = x - 18$

$x^2 - 11x + 18 = 0$

$(x - 2)(x - 9) = 0$

$x = 2, 9$

# Exponential Growth and Decay

## Exponential Functions: $y = b^x$

| | |
|---|---|
| **Exponential Growth**<br><br>$y = b^x$<br><br>$b > 1$ | 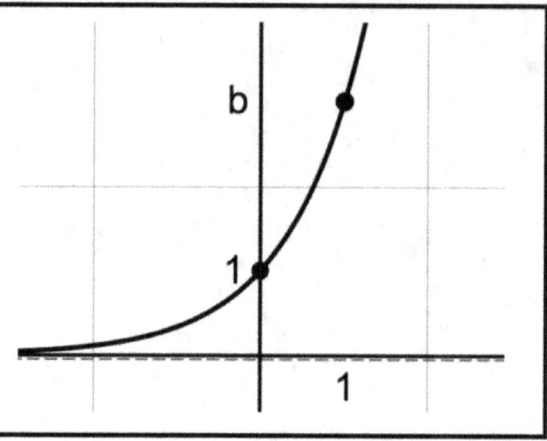 |
| **Exponential Decay**<br><br>$y = b^x$<br><br>$b < 1$ | 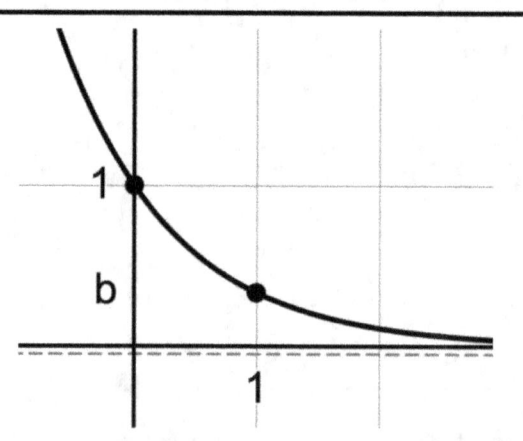 |

| **Reference Points:** | $(0, 1)$ & $(1, b)$ |
|---|---|

# Exponential Functions: $y = ab^x$

| Growth of Money | Compounded Yearly | $A = P(1 + r)^t$ |
|---|---|---|
| | Compounded $n$ times per year | $A = P\left(1 + \dfrac{r}{n}\right)^{nt}$ |
| | Compounded Continuously | $A = Pe^{rt}$ |

| Decay of a Value | Depreciates at a yearly rate, $r$ | $A = A_0(1 - r)^t$ |
|---|---|---|
| | Depreciates by half every $h$ years | $A = A_0\left(\dfrac{1}{2}\right)^{\frac{t}{h}}$ |

| Terms | $t$ = Time (in years) <br> $r$ = rate (yearly rate) <br> $A$ = Amount at time $t$ <br> $P$ = Principal (Initial Value) <br> $A_0$ = Initial Value |
|---|---|

| Exponential Growth – Ex. 1 |
|---|

If $80,000 is invested in a bank, find the value of the investment under the following conditions.

| Conditions | Result |
|---|---|
| 3% interest, Compounded annually, For 50 years. | $A = P(1+r)^t$ <br> $A = 80{,}000(1+0.03)^{50}$ <br> $A = 80{,}000(1.03)^{50}$ <br> $A = \$350{,}712.50$ |
| 3% interest, Compounded monthly, For 50 years. | $A = P\left(1+\dfrac{r}{n}\right)^{nt}$ <br> $A = 80{,}000\left(1+\dfrac{.03}{12}\right)^{12(50)}$ <br> $A = 80{,}000(1.025)^{600}$ <br> $A = \$350{,}864.60$ |
| 3% interest, compounded continuously, For 50 years. | $A = Pe^{rt}$ <br> $A = (80{,}000)\,e^{(.03)(50)}$ <br> $A = (80{,}000)\,e^{1.5}$ <br> $A = \$358{,}535.10$ |

| | **Exponential Decay – Ex. 2** |
|---|---|
| \multicolumn{2}{l}{Joe bought a new car for $60,000. If the value of the car depreciates at a rate of 6% per year, how much will his car be worth in ten years? In 15 years? When will his car be worth $10,000 ?} |
| Value of car in 10 years | $A = A_0(1-r)^t$ <br> $A = 60,000\,(1-.06)^{10}$ <br> $A = 60,000\,(0.94)^{10}$ <br> $A = \$32,316.90$ |
| Value of car in 15 years | $A = A_0(1-r)^t$ <br> $A = 60,000\,(1-.06)^{15}$ <br> $A = 60,000\,(0.94)^{15}$ <br> $A = \$23,717.50$ |
| When car is worth $10,000 | $A = A_0(1-r)^t$ <br> $10,000 = 60,000\,(1-.06)^t$ <br> $0.16667 = (0.94)^t$ <br> $\log(0.16667) = \log(0.94)^t$ <br> $\log(0.16667) = t \cdot \log(0.94)$ <br> $t = \dfrac{\log(0.16667)}{\log(0.94)} \approx 29\ yrs.$ |

## Exponential Growth (Unknown Rate) – Ex. 3

Sally, the scientist, is growing a new kind of algae in a laboratory. She started with 1000 cells. In 1 hour, there were 1234 algae cells. How many algae cells will there be in 5 hours, after she started?
How long will it take the number of algae to triple?

| | |
|---|---|
| Determine the growth rate. | $r = \dfrac{Change}{Original} = \dfrac{new - old}{old}$ <br><br> $r = \dfrac{1234 - 1000}{1000} = \dfrac{234}{1000} = .234$ <br><br> $r = 23.4\%$ increase per hour |
| Quantity in 5 hours | $A = A_0(1+r)^t$ <br><br> $A = 1000\,(1 + .234)^5$ <br><br> $A = 1000\,(1.234)^5$ <br><br> $A = 2861.38 \approx 2861$ algae |
| When number of algae triples | $A = A_0(1+r)^t$ <br><br> $3000 = 1000\,(1 + .234)^t$ <br><br> $3 = (1.234)^t$ <br><br> $\log(3) = \log 1.234^t$ <br><br> $\log(3) = t \cdot \log(1.234)$ <br><br> $t = \dfrac{\log(3)}{\log(1.234)} \approx 5.2\ hours$ |

| | Half Life – Ex. 4 |
|---|---|
| | Pd-100 has a half-life of 3.6 days. Sam, the scientist, started with one mole of this substance.<br>Note: $1\ mole = 6.023 \times 10^{23}\ atoms$<br><br>(1) How many atoms will be present after 15 days?<br>(2) When will the number of atoms be 6 million? |
| Quantity after 15 days | $A = A_0 \left(\frac{1}{2}\right)^{\frac{t}{h}}$<br><br>$A = 6.023 \times 10^{23} \left(\frac{1}{2}\right)^{\frac{15}{3.6}}$<br><br>$A = 3.35\ E\ 22 = 3.35 \times 10^{22}\ atoms$ |
| When 6 million atoms remain. | $A = A_0 \left(\frac{1}{2}\right)^{\frac{t}{h}}$<br><br>$6{,}000{,}000 = 6.023 \times 10^{23} \left(\frac{1}{2}\right)^{\frac{t}{3.6}}$<br><br>$9.962 \times 10^{-18} = (.5)^{\frac{t}{3.6}}$<br><br>$\log(9.962 \times 10^{-18}) = \left(\frac{t}{3.6}\right) \log(.5)$<br><br>$\frac{t}{3.6} = \frac{\log(9.962 \times 10^{-18})}{\log(0.5)} = 56.48$<br><br>$t = (56.48)(3.6) \approx 203\ days$ |

# Regression

| |
|---|
| **Regression** |
| **Regression**, in math, is fitting a curve, $y = f(x)$, to a set of points. Mathematical techniques can be used, but are not discussed in this section.<br><br>Here, regression will be demonstrated in two ways:<br>   1) With the Desmos, online graphing tool, and<br>   2) With a TI graphing calculator.<br><br>After that, regression will be used to find the time until the next eruption of the "Old Faithful" geyser at Yellowstone National Park. |
| Type of Regression should match the data. Graph the data to see what type of regression will be best.<br>- Linear Regression        (straight line)<br>- Quadratic Regression   (parabolic shape)<br>- Exponential Regression (exponential shape)<br>- Etc. |
| Goodness of Fit:<br>- $R$ and $R^2$ values tell how well the regression matches the set of points.<br>- The closer $\lvert R \rvert$ and $R^2$ are to 1, the better.<br>- If $R^2 = 1$ it's a perfect fit. |

## Regression on Desmos (Desmos.com)

Regression on Desmos can be done on:
- Using the Desmos app, on your mobile phone.
- Using the Desmos website, on you laptop.

The following set of points will be used to demonstrate regression on the Desmos website.

| x | 0 | 1 | 2 | 3 | 4 | 5 | 6 |
|---|---|---|---|---|---|---|---|
| y | 6 | 2 | −.5 | −2.5 | 0 | 2 | 8 |

| | |
|---|---|
| • Click the + sign, top left.<br>• Select "Table"<br>• Input all data points | $x_1$ \| $y_1$<br>0 \| 6<br>1 \| 2<br>2 \| −.5<br>3 \| −2.5 |
| Adjust graph settings to fit your data. Click on wrench. | 🔧 |
| Plot the points.<br>Type this: (x1, y1)<br>Desmos formats subscripts. | $(x_1, y_1)$ |

| **Regression on Desmos (Continued)** ||
|---|---|
| Looks like quadratic regression is appropriate. | 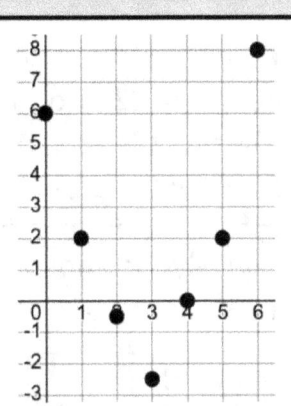 |
| Type this: $y_1 \sim ax_1^2 + bx_1 + c$ <br><br> Notes: <br> • Just type y1 to get $y_1$ <br> • Just type x1 to get $x_1$ <br> • Use " ~ " (not " = ") |  $y_1 \sim ax_1^2 + bx_1 + c$ <br><br> STATISTICS   RESIDUALS <br> $R^2 = 0.9785$   $e_1$  plot <br><br> PARAMETERS <br> $a = 0.970238$ <br> $b = -5.58929$ <br> $c = 6.29762$ |
| Graph shows regression curve and data points. | 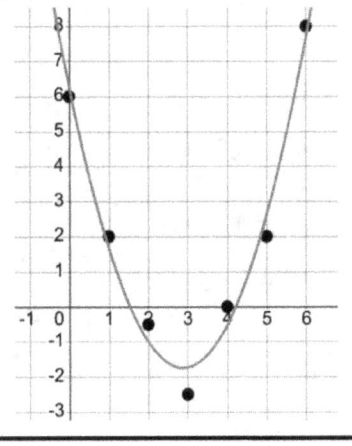 |

## Regression on Desmos -- Using the Equation

Use the quadratic regression to find " $y$ " when $x = 12$. Do this in two ways:
1) Using the equation
2) Using the graph

Previously, Desmos was used to fit a quadratic function $y = .97x^2 - 5.59x + 6.30$ to these points:

| x | 0 | 1 | 2 | 3 | 4 | 5 | 6 |
|---|---|---|---|---|---|---|---|
| y | 6 | 2 | −.5 | −2.5 | 0 | 2 | 8 |

| | |
|---|---|
| Find " $y$ " when $x = 12$ Use equation. | $y = .97x^2 - 5.59x + 6.30$ <br> $y = .97(12)^2 - 5.59(12) + 6.30$ <br> $y = 78.90$ |
| Find " $y$ " when $x = 12$ Use graph. <br><br> Graph on next page. | • Enter equation: <br>    $y = .97x^2 - 5.59x + 6.30$ <br> • Enter equation: $x = 12$ <br> • Expand window to show intersection. <br> • Click on intersection to shop point $(12, 78.9)$. |

## Regression on Desmos -- Using the Equation

### Graph

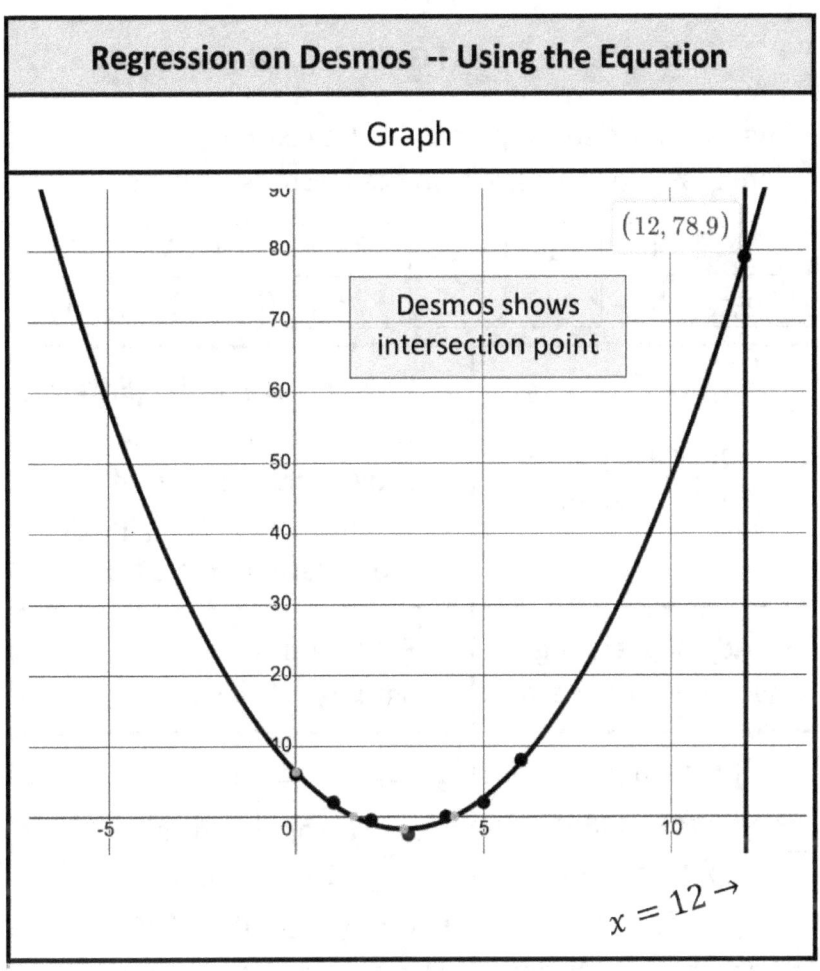

(12, 78.9)

Desmos shows intersection point

$x = 12 \rightarrow$

| **Regression on TI Graphing Calculator** |

The following set of points will be used to demonstrate regression on the Desmos website.

| x | 0 | 1 | 2 | 3 | 4 | 5 | 6 |
|---|---|---|---|---|---|---|---|
| y | 6 | 2 | −.5 | −2.5 | 0 | 2 | 8 |

| | |
|---|---|
| Input data points into a table. | • Press **[ stat ]**, EDIT, **[Enter]**<br>• Clear data from L1 & L2<br>• Move cursor to top of column and press **[clear]**.<br>• Input data under L1 & L2. |
| Adjust graphing window to fit data. | • Press **[window]**<br>• Set x & y. (max & min) |
| Plot data points. Look at shape to determine type of regression | • Press **[ y= ]** button<br>• Use up arrow to select **Plot1** to plot points.<br>• Press **[graph]** button. |
| Points are plotted. Looks like Quadratic regression Is appropriate. | 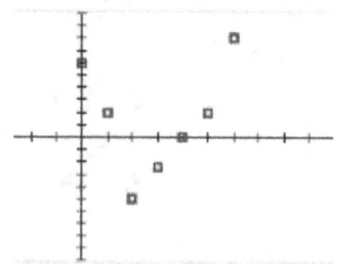 |

| Regression on TI Graphing Calculator | |
|---|---|
| Continued … | |
| Turn on diagnostics to get $R$ & $R^2$ Values | • Press [ 2$^{nd}$ ] , [ 0 ] to select **Catalog** .<br>• Scroll down and select "DiagnosticsOn"<br>• ENTER, ENTER, Done. |
| Do regression on your data. | • Press [ stat ] , CALC.<br>• Scroll down to select "QuadReg", [enter] |
| Select regression information (No Graph) | • Take all defaults.<br>• Scroll down to Calculate. |
| Select regression information (With Y1 Graph) | • Scroll to "Store RegEQ:"<br>• [ vars ], Y-VARS , Function<br>• [enter] , Y1, [enter]<br>• Scroll down to Calculate |
| Screen should look like this. | QuadReg<br>Xlist:L₁<br>Ylist:L₂<br>FreqList:<br>Store RegEQ:Y₁ |

| Regression on TI Graphing Calculator | |
|---|---|
| Continued … | |
| Regression Information is given | QuadReg<br>y=ax²+bx+c<br>a=1.130952381<br>b=-6.392857143<br>c=5.976190476<br>R²=0.9049546944 |
| Graph the Regression equation | • Press [ y= ], [graph] |
| The Graph and Points | |

To connect your TI-84 graphing calculator to your computer, install the free TI Connect software application via "education.ti.com/ticonnect" on your computer.

## Regression on TI Graphing Calc. -- Using the Equation

Use the quadratic regression to find " $y$ " when $x = 12$. Do this in two ways:
1) Using the equation
2) Using the graph

Previously, Desmos was used to fit a quadratic function $y = 1.1x^2 - 6.4x + 6$ to these points:

| x | 0 | 1 | 2 | 3 | 4 | 5 | 6 |
|---|---|---|---|---|---|---|---|
| y | 6 | 2 | −.5 | −2.5 | 0 | 2 | 8 |

| | |
|---|---|
| Find " $y$ " when $x = 12$ Use equation. | $y = 1.13x^2 - 6.39x + 5.98$ <br> $y = 1.13(12)^2 - 6.39(12) + 5.98$ <br> $y = 92.02$ |
| Find " $y$ " when $x = 1$ Use graph. <br><br> Graph on next page. | • Press **[window]**, expand window <br> • Press **[graph]** <br> • Press **[trace]** <br> • Up arrow toggles between the plot (points) and Y1 (function) Selection displayed along top <br> • Press up arrow to select Y1 <br> • Type: 12 |

**Regression on TI Graphing Calc. -- Using the Equation**

Graph

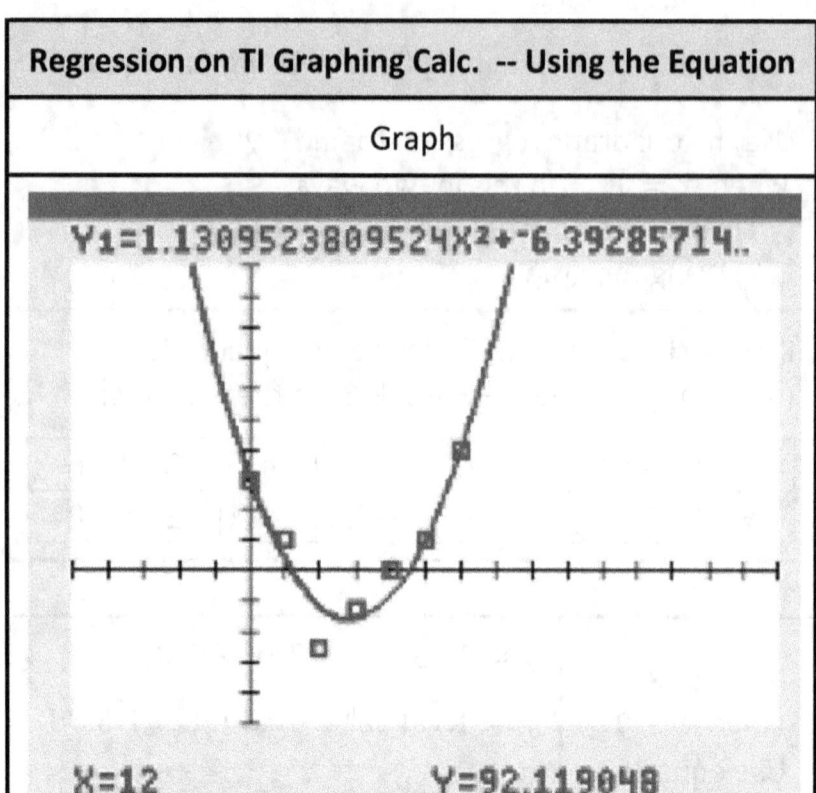

When in **[trace]** mode, just type the value of X
And the value of Y will be displayed.

---

To connect your TI-84 graphing calculator to your computer, install the free TI Connect software application via "education.ti.com/ticonnect" on your computer.

## Old Faithful Geyser – Problem Statement

True story: My family's tour, at Yellowstone National Park, included a stop at the Old Faithful geyser. The Old Faithful geyser is famous because it erupts on a relatively predictable schedule. On average, the geyser erupts for about 3.5 minutes. Then, the next eruption occurs about 71 minutes later. In general, if the geyser eruption is short, then the time until the next eruption is also shorter and vice-versa.

Our tour-guide made sure we stopped at Old Faithful before the next predicted eruption.

I decided to look online to find some data about the Old Faithful geyser and use regression to develop my own predictor.

Problem Statement: Find and use online data for the Old Faithful geyser to predict the time until the next eruption.

## Old Faithful Geyser Regression – Getting Data

Find and use online data for the Old Faithful geyser to predict the time until the next eruption.

---

There are many sources of online data. I just picked one that looked reasonable and reliable. I found some data from Carnegie Mellon University (CMU) that included 273 records of:
- Eruption Time
- Time to Next Eruption

| Data was copied from online and pasted into an Excel Spreadsheet. The first few records are shown. | A272     $f_x$   2: <br><br> A <br> 1   1 , 3.600 , 79 <br> 2   2 , 1.800 , 54 <br> 3   3 , 3.333 , 74 <br> 4   4 , 2.283 , 62 |
|---|---|
| Within Excel, data was separated and formatted. If necessary, search online for how to do this. |     A     B     C <br> 1   1   3.6   79 <br> 2   2   1.8   54 <br> 3   3   3.333   74 <br> 4   4   2.283   62 <br> 5   5   4.533   85 |

## Old Faithful Geyser Regression – Formatting Data

Find and use online data for the Old Faithful geyser to predict the time until the next eruption.

Make sure you have all the data you need. What is your output? What are your input(s)?

| | |
|---|---|
| Output | D. Time until next eruption. |
| Input(s) | The input could be the time of the last eruption. Or, perhaps the prediction would be better if more inputs were used. These inputs were considered:<br>A. Previous duration of eruption<br>B. Previous time until next eruption.<br>C. Duration of latest eruption |

For these Values, Averages were used.

| | A | B | C | D |
|---|---|---|---|---|
| | Dur_Prev | Next_Prev | Duration | Next |
| 2 | 3.475 | 71.000 | 3.6 | 79 |
| 3 | 3.600 | 79.000 | 1.8 | 54 |
| 4 | 1.800 | 54.000 | 3.333 | 74 |
| 5 | 3.333 | 74.000 | 2.283 | 62 |
| 6 | 2.283 | 62.000 | 4.533 | 85 |

**Old Faithful Geyser Regression – Load Desmos Table**

Find and use online data for the Old Faithful geyser to predict the time until the next eruption.

Go to Desmos.com and click on the "Graphing" tools. Click on the "+" then "f(x) expression."

Note: "Table" is for 2-columns. "f(x) expression" is for multiple columns.

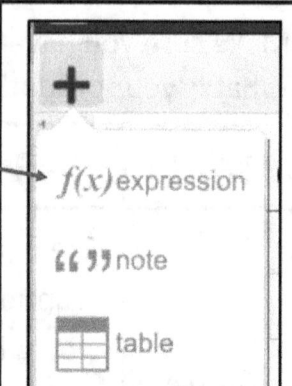

Copy first 4 columns in Excel. Paste into Desmos.

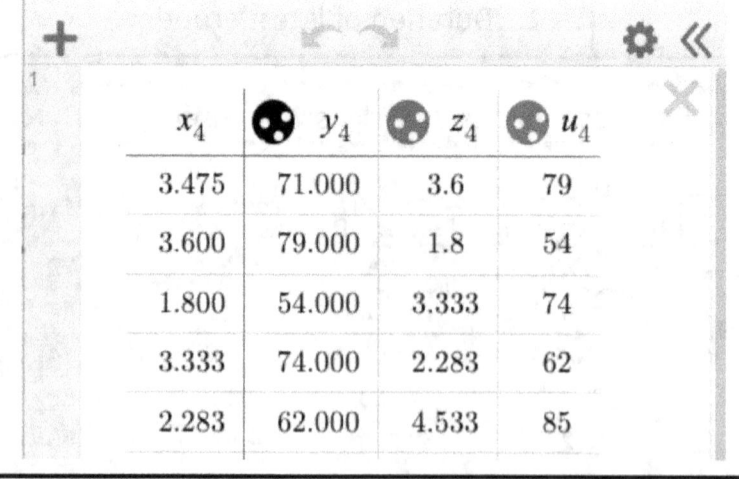

| $x_4$ | $y_4$ | $z_4$ | $u_4$ |
|---|---|---|---|
| 3.475 | 71.000 | 3.6 | 79 |
| 3.600 | 79.000 | 1.8 | 54 |
| 1.800 | 54.000 | 3.333 | 74 |
| 3.333 | 74.000 | 2.283 | 62 |
| 2.283 | 62.000 | 4.533 | 85 |

## Old Faithful Geyser Regression – Do Regression

| Prev Dur $x_4$ | Prev Next $y_4$ | Dur $z_4$ | Next $u_4$ |
|---|---|---|---|
| 3.475 | 71.000 | 3.6 | 79 |
| 3.600 | 79.000 | 1.8 | 54 |
| 1.800 | 54.000 | 3.333 | 74 |

Two regressions were done to predict "Next" eruption

$u_4 \sim az_4 + b$

STATISTICS     RESIDUALS

$r^2 = 0.8115$  $e_1$
$r = 0.9008$

PARAMETERS
$a = 10.7296$   $b = 33.4744$

$u_4 \sim ax_4 + by_4 + cz_4 + d$

STATISTICS     RESIDUALS

$R^2 = 0.8263$  $e_4$  plot

PARAMETERS
$a = 3.11438$   $b = -0.282415$   $c = 10.5745$
$d = 43.1838$

| **Old Faithful Geyser Regression – Conclusion** |
|---|
| Find and use online data for the Old Faithful geyser to predict the time until the next eruption. |

| | |
|---|---|
| Compare regression with 1 input to the regression with 3 inputs. | The regression with 1 input is more practical and had $r^2 = 0.81$ <br><br> The regression with 3 inputs is more complex and had $R^2 = 0.83$ (not much better) |
| Conclusion | The regression with 1 input is more practical and simple to use. <br><br> The additional accuracy of the 3-input regression is not worth the trouble! |
| Regression based on one input: Duration of most recent eruption. | $y = 10.7x + 33.5$ <br><br> $y$ = Time to next eruption. <br> $x$ = Duration of last eruption. |

# Systems of Linear Equations

## Systems of Linear Equations in 2 Variables

A **System of Linear Equations** is a set of linear equations in the same two variables. When two linear systems are graphed, there are three possibilities, as shown below.

| Possibility | Example | Graph |
|---|---|---|
| Intersecting Lines<br><br>1 Solution | $y = x - 3$<br><br>$y = -x + 5$ | 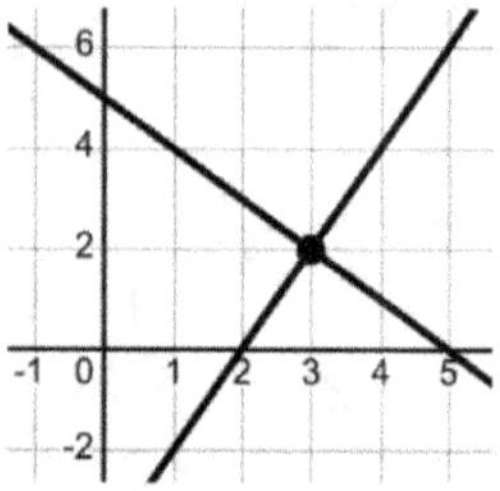 |
| Parallel Lines<br><br>0 Solutions | $y = \frac{x}{2} + 3$<br><br>$y = \frac{x}{2} - 2$ | 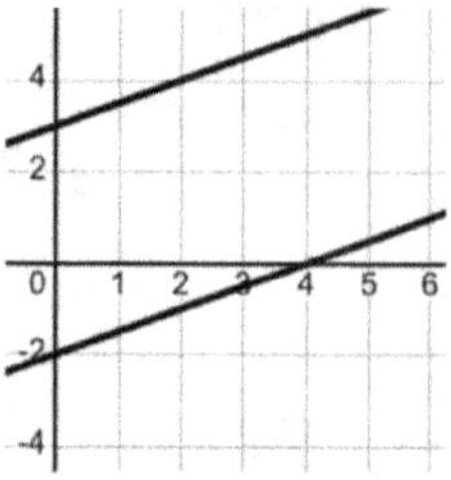 |
| Same Lines<br><br>∞ Solutions | $y - 2 = x$<br><br>$y = x + 2$ | 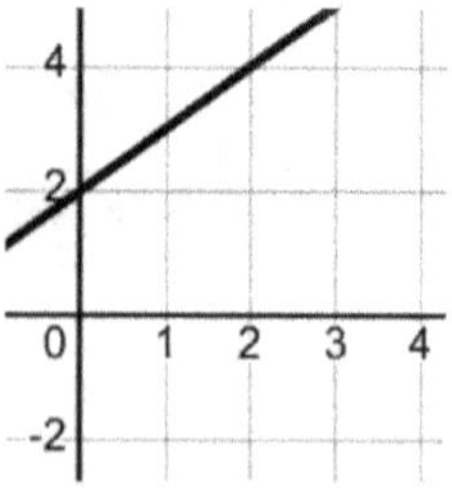 |

## Systems of Linear Equations in 2 Variables
## How to Solve

### Algebraic Solutions

| Substitution | Elimination |
|---|---|
| • Isolate one variable<br>• Substitute it into the other equation.<br>• That equation will have only one variable.<br>• Solve it to get one variable.<br>• Use known variable to find the other. | • Multiply equation(s) so one variable has the same coefficient in both equations.<br>• Add both equations to eliminate one variable.<br>• Solve it to get one variable.<br>• Use known variable to find the other. |

### Graphical Solution

- Graph both equations on the same coordinate axis.
- Find the point where they intersect.
- The coordinates of the intersection is the solution.

## Systems of Linear Equations – Ex. 1a

Given this system: $\begin{cases} x - 2y = 5 \\ 4x + 3y = 9 \end{cases}$

Solve using substitution, elimination, and graphing.

### Substitution Solution

| | |
|---|---|
| Isolate one variable | $x - 2y = 5$ <br> $x = 5 + 2y$ |
| Substitute into other equation and solve. | $4x + 3y = 9$ <br> $4(5 + 2y) + 3y = 9$ <br> $20 + 8y + 3y = 9$ <br> $11y = -11$ <br> $y = -1$ |
| Use known variable to find the other. | $x - 2y = 5$ <br> $x - 2(-1) = 5$ <br> $x + 2 = 5$ <br> $x = 3$ |
| Solution | $(x, y) = (3, -1)$ |

## Systems of Linear Equations – Ex. 1b

Given this system: $\begin{cases} x - 2y = 5 \\ 4x + 3y = 9 \end{cases}$

Solve using substitution, elimination, and graphing.

### Elimination Solution

| | |
|---|---|
| Multiply eqns. So one variable has matching coefficients. | $x - 2y = 5$ $\quad\to\quad$ $4x - 8y = 20$ <br> $4x + 3y = 9$ $\qquad\quad$ $4x + 3y = 9$ |
| Subtract 2nd from 1st eqn. and solve for one variable. | $4x - 8y = 20$ <br> $-(\,4x + 3y = 9\,)$ <br> $\overline{\phantom{xxxx}-11y = 11}$ <br> $y = -1$ |
| Use known variable to find the other. | $x - 2y = 5$ <br> $x - 2(-1) = 5$ <br> $x + 2 = 5$ <br> $x = 3$ |
| Solution | $(x, y) = (3, -1)$ |

## Systems of Linear Equations – Ex. 1c

Given this system: $\begin{cases} x - 2y = 5 \\ 4x + 3y = 9 \end{cases}$

Solve using substitution, elimination, and graphing.

### Graphing Solution

| $x - 2y = 5$ | $4x + 3y = 9$ |
|---|---|
| $x = 0 \rightarrow y = -\frac{5}{2}$ | $x = 0 \rightarrow y = 3$ |
| $y = 0 \rightarrow x = 5$ | $y = 0 \rightarrow x = \frac{9}{4}$ |

Intersection: $(x, y) = (3, -1)$

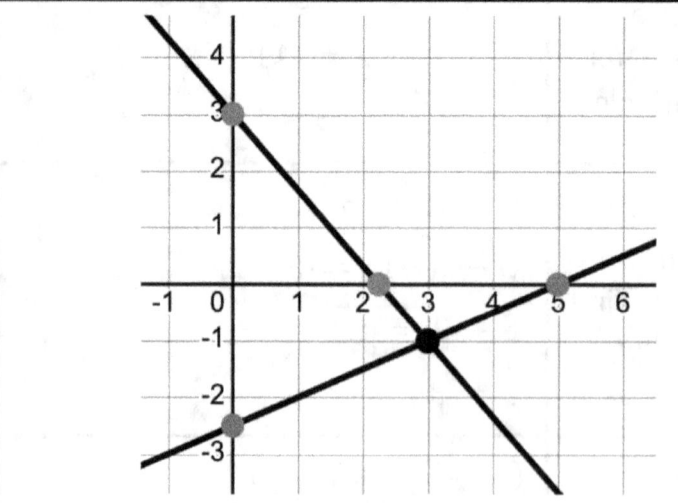

# Systems of Linear Inequalities

## Systems of Linear Inequalities in 2 Variables

A **System of Linear Inequalities** is a set of linear inequalities in the same two variables.

To solve a system of linear inequalities, do this:
- Graph both on the same coordinate axis.
- Shade one side of each to show the feasible region.
- The solution is the set of ordered pairs, within the intersection of the shaded regions.

| Examples |||
|---|---|---|
| $y > x + 1$ | $y \leq -x + 3$ | $y > x + 1$ AND $y \leq -x + 3$ |

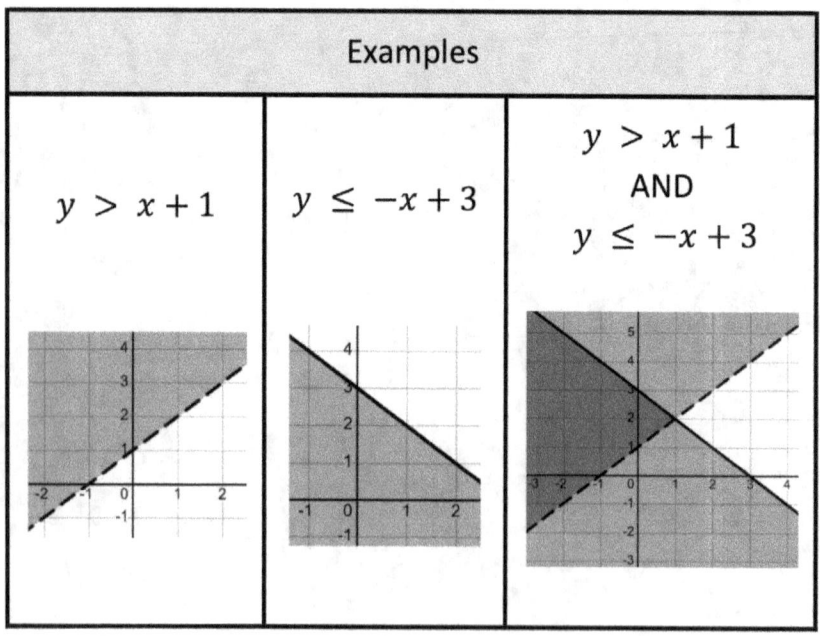

## Systems of Linear Inequalities in 2 Variables
### Ex. 1a

Given this system of inequalities:

$$y < -\frac{1}{2}x + 4$$

$$y \geq x - 2$$

1) Graph the inequalities to show the feasible region.

2) Determine if the following points are in the feasible region: $(1,1), (2,3), (3,1), (4,1), (6,1), (6,4)$

Graph

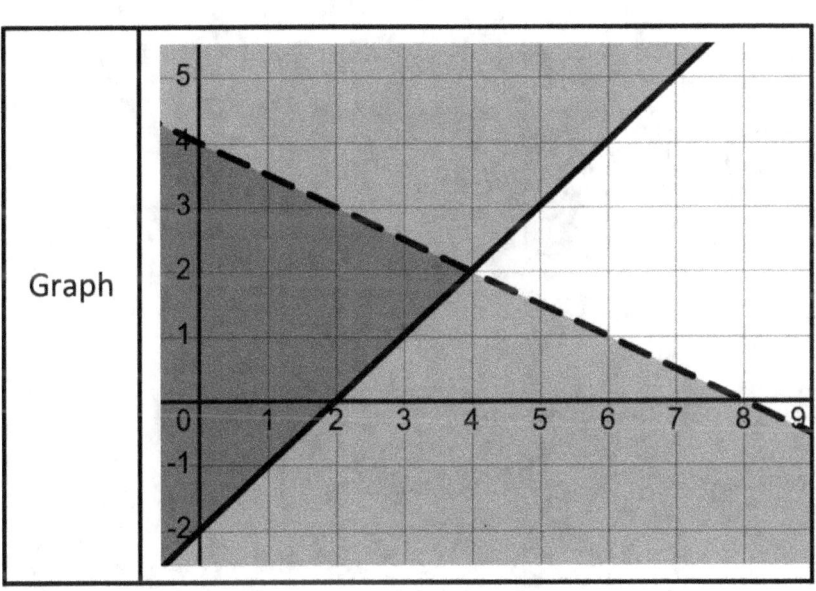

## Systems of Linear Inequalities in 2 Variables
### Ex. 1b

Given this system of inequalities:
$$y < -\frac{1}{2}x + 4$$
$$y \geq x - 2$$

1) Graph the inequalities to show the feasible region.

2) Determine if the following points are in the feasible region: $(1,1), (2,3), (3,1), (4,1), (6,1), (6,4)$

$(1, 1)$  Yes

$(2, 3)$  No

$(3, 1)$  Yes

$(4, 1)$  No

$(6, 1)$  No

$(6, 4)$  No

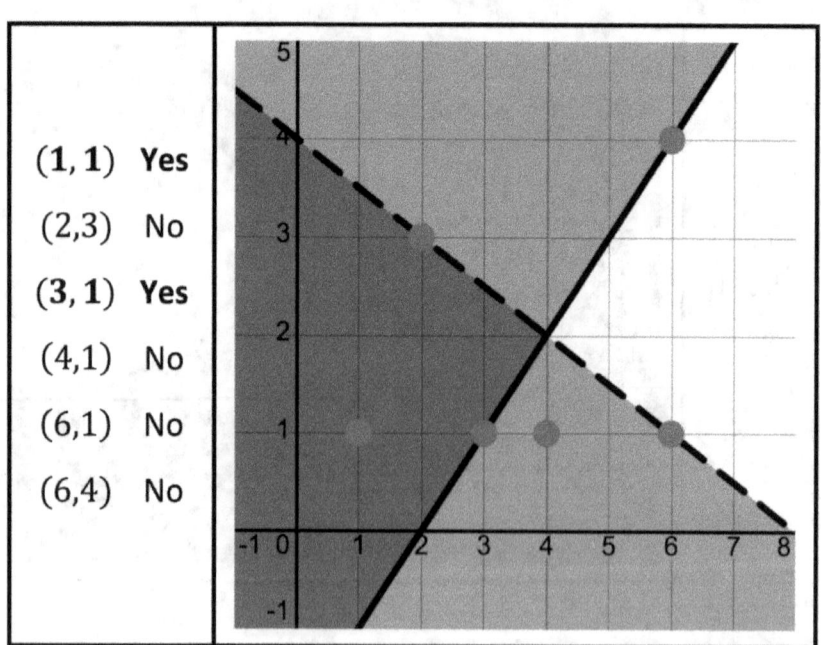

# Solving Inequalities

## Solving Inequalities
## With the Critical Value Method (CVM)

The **Critical Value Method (CVM)** is a process used to solve any inequality – simple or complex. The process is outlined below.

| | |
|---|---|
| 1. | Identify restrictions on $x$, if any. |
| 2. | Find critical values.<br>• Change the inequality sign to an equal sign. The solutions are critical values.<br>• Restrictions on $x$ (if any) are also critical values. |
| 3. | Sketch a number line, with the critical values.<br>Note: $n$ critical values will create $(n+1)$ intervals on the number line. |
| 4. | Test one value, in each interval, using the original inequality. Do not use boundary numbers.<br>• If it makes the original inequality true, then all numbers in that interval are solutions.<br>• If it makes the original inequality false, then all numbers in that interval are NOT solutions. |
| 5. | Solution: Write solution in interval notation. |

| Critical Value Method (CVM) -- Ex. 1 ||
|---|---|
| Given: $\|x+2\| \leq 5$ | Solve for $x$ |
| 1. Identify restrictions on $x$ | None. |
| 2. Find Critical Values<br><br>Change the inequality sign to an equal sign and solve. | $\|x+2\| = 5$ |
| | $x + 2 = 5$<br>$x = 3$ |
| | $x + 2 = -5$<br>$x = -7$ |
| | $x = -7, 3$  (Critical Values) |
| 3. Sketch a number line, with the critical values. | ●————● <br> -7        3 |
| 4. Test one value, in each interval, using the original inequality.<br><br>$\|x+2\| \leq 5$ | F●  T  ●F<br>  -7      3<br><br>$x = -8 \;\rightarrow\;$ False<br>$x = \phantom{-}0 \;\rightarrow\;$ True<br>$x = \phantom{-}5 \;\rightarrow\;$ False |
| 5. Solution: | $x = [-7, 3]$ |

## Critical Value Method (CVM) -- Ex. 2

Given: $|x + 2| \geq 5$    Solve for $x$

| | | |
|---|---|---|
| 1. | Identify restrictions on $x$ | None. |
| 2. | Find Critical Values.<br><br>Change the inequality sign to an equal sign and solve. | $|x + 2| = 5$ |
| | | $x + 2 = 5$<br>$x = 3$ |
| | | $x + 2 = -5$<br>$x = -7$ |
| | | $x = -7, 3$   (Critical Values) |
| 3. | Sketch a number line, with the critical values. | ●————————●<br>-7            3 |
| 4. | Test one value, in each interval, using the original inequality.<br><br>$\|x + 2\| \geq 5$ | T   F   T<br>——●———●——<br>  -7    3<br><br>$x = -8 \rightarrow$ True<br>$x = \phantom{-}0 \rightarrow$ False<br>$x = \phantom{-}5 \rightarrow$ True |
| 5. | Solution: | $x = (-\infty, -7] \cup [3, \infty)$ |

| | **Critical Value Method (CVM) -- Ex. 3** | |
|---|---|---|
| | Given: $\sqrt{x+2} < 5$ | Solve for $x$. |
| 1. | Identify restrictions on $x$ | $x + 2 \geq 0$ <br> $x \geq -2$ |
| 2. | Find Critical Values. <br><br> Change inequality sign to an equal sign and solve. | $\sqrt{x+2} = 5$ <br> $x + 2 = 5^2$ <br> $x = 23 \quad$ (Critical Value) |
| 3. | Sketch a number line, with the critical values. | —————o————— <br> $\phantom{xxxxx}23$ |
| 4. | Test one value, in each interval, using the original inequality. <br><br> $\sqrt{x+2} < 5$ | $\phantom{xx}$T $\phantom{x}$ F <br> ———o——— <br> $\phantom{xxx}23$ <br> $x = 22 \rightarrow$ True <br> $x = 24 \rightarrow$ False |
| 5. | Solution: <br> Must consider restrictions on $x$. | $x = [-2, 23)$ |

| | **Critical Value Method (CVM) -- Ex. 4** | |
|---|---|---|
| | Given: $x^2 + 7x > 0$ | Solve for $x$. |
| 1. | Identify restrictions on $x$ | None. |
| 2. | Find Critical Values. Change inequality sign to an equal sign and solve. | $x^2 + 7x = 0$ <br> $x(x+7) = 0$ <br> $x = 0, -7$ |
| 3. | Sketch a number line, with the critical values. | ──○────○── <br>   -7    0 |
| 4. | Test one value, in each interval, using the original inequality. <br><br> $x^2 + 7x > 0$ | T   F   T <br> ──○───○── <br>   -7    0 <br><br> $x = -8 \rightarrow$ True <br> $x = -4 \rightarrow$ False <br> $x = 2 \rightarrow$ True |
| 5. | Solution: | $x = (-\infty, -7) \cup (0, \infty)$ |

## Critical Value Method (CVM) -- Ex. 5

Given: $\dfrac{x-4}{x-1} \leq 0$   Solve for $x$.

| | | |
|---|---|---|
| 1. | Identify restrictions on $x$ | $x \neq 1$   (Critical Value) |
| 2. | Find Critical Values.<br><br>Change inequality sign to an equal sign and solve. | $\dfrac{x-4}{x-1} = 0$<br><br>$x - 4 = 0$<br><br>$x = 4$   (Critical Value) |
| 3. | Sketch a number line, with the critical values. | (number line with open circle at 1 and closed circle at 4) |
| 4. | Test one value, in each interval, using the original inequality.<br><br>$\dfrac{x-4}{x-1} \leq 0$ | (number line marked F, T, F)<br><br>$x = 0 \rightarrow$ False<br>$x = 2 \rightarrow$ True<br>$x = 5 \rightarrow$ False |
| 5. | Solution: | $x = (1, 4]$ |

| | **Critical Value Method (CVM) -- Ex. 6** | |
|---|---|---|
| | Given: $\dfrac{3x+1}{x-2} \geq 4$ | Solve for $x$. |
| 1. | Identify restrictions on $x$ | $x \neq 2$     (Critical Value) |
| 2. | Find Critical Values. Change inequality sign to an equal sign and solve. | $\dfrac{3x+1}{x-2} = 4$ <br> $3x + 1 = 4x - 8$ <br> $x = 9$     (Critical Value) |
| 3. | Sketch a number line, with the critical values. | ──────○────●────── <br>        2      9 |
| 4. | Test one value, in each interval, using the original inequality. $\dfrac{3x+1}{x-2} \geq 4$ |   F    T    F <br> ──────○────●────── <br>       2     9 <br> $x = 0 \rightarrow$ False <br> $x = 3 \rightarrow$ True <br> $x = 10 \rightarrow$ False |
| 5. | Solution: | $x = (\,2, 9\,]$ |

# Trig Topics - Part 1

# Right Triangles

## Right Triangles

Right Triangles have one angle = 90°

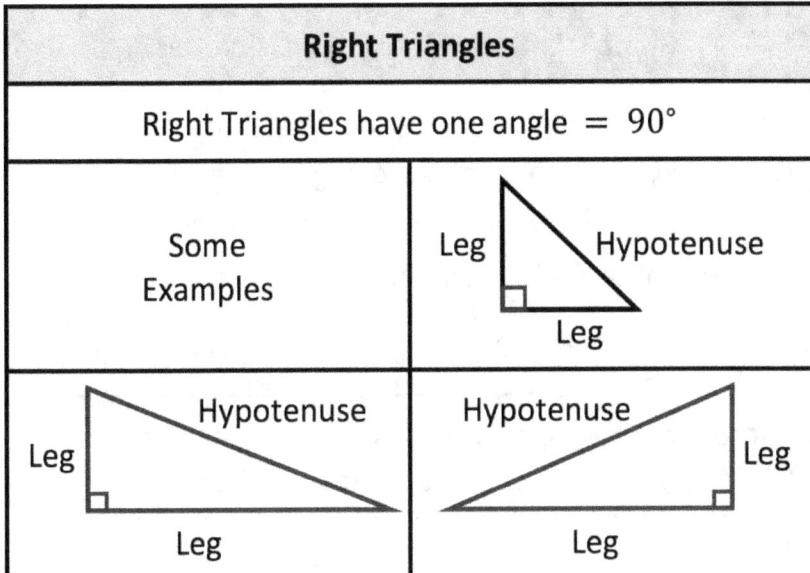

Some Examples

In a right triangle, the side opposite the right angle is the hypotenuse. The sides adjacent (next to) the right angle are often called the "legs."

The Hypotenuse is always the largest of the 3 sides.

## Pythagorean Theorem – For Right Triangles

If the length of the hypotenuse, of a right triangle, is $c$ and the lengths of the other two sides (legs) are $a$ and $b$, then:

$$c^2 = a^2 + b^2$$

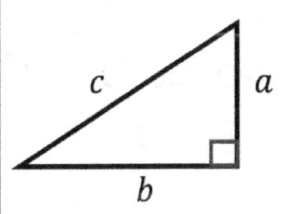

## Right Triangles – Infinite Possibilities!

The other two angles, in a right triangle must add up to 90° because the sum of the angles, in any triangle, is 180°. There are infinite possibilities with the measure of the other two angles. For example:

90, 45, 45   or   90, 20, 70   or   90, 19.2, 70.8 ...

There are three right triangles that are used often. So, it's a good idea to be familiar with them.

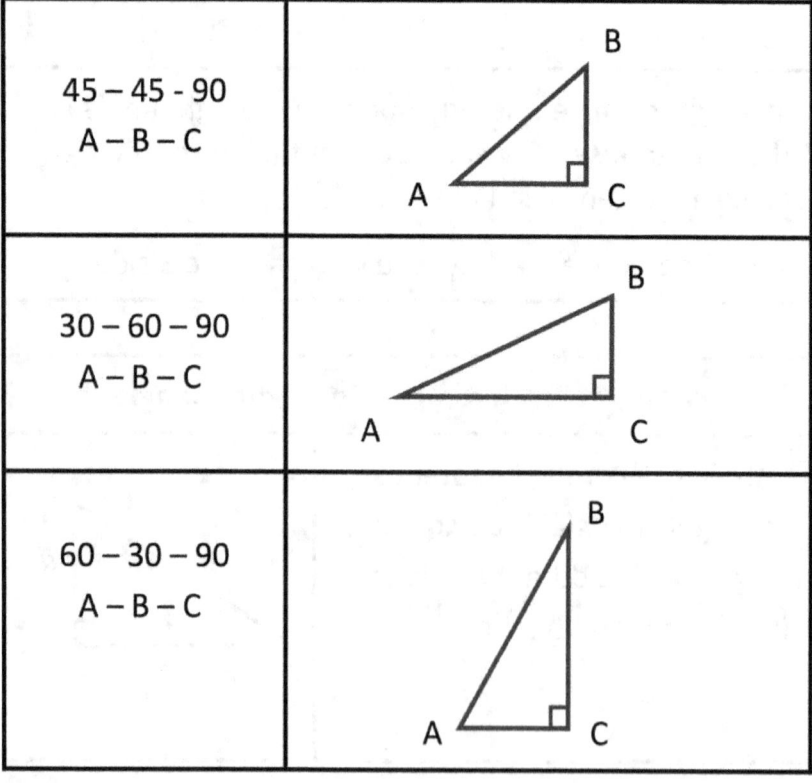

| 45 – 45 - 90  A – B – C | |
| 30 – 60 – 90  A – B – C | |
| 60 – 30 – 90  A – B – C | |

## 3 "Popular" Right Triangles, Hypotenuse = 1

If the hypotenuse = 1 then the other two lengths are:

| | |
|---|---|
| 45 – 45 - 90<br>A – B – C | 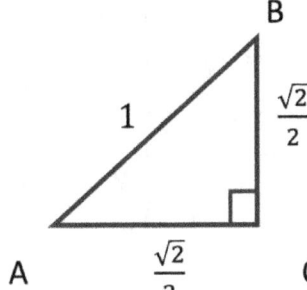 |
| 30 – 60 – 90<br>A – B – C | 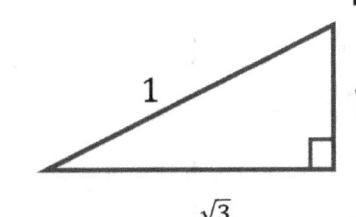 |
| 60 – 30 – 90<br>A – B – C | 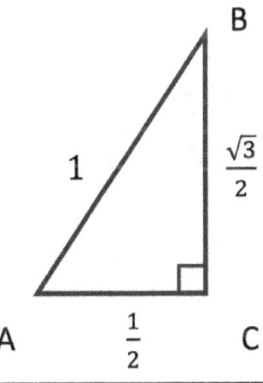 |

## Right Triangles – 3 "Popular" Triangles

If the hypotenuse = $h$ then the other two lengths are:

| | |
|---|---|
| 45 – 45 - 90<br>A – B – C | Hypotenuse $h$; leg opposite B (AC) = $h \cdot \frac{\sqrt{2}}{2}$; leg BC = $h \cdot \frac{\sqrt{2}}{2}$ |
| 30 – 60 – 90<br>A – B – C | Hypotenuse $h$ (AB); AC = $h \cdot \frac{\sqrt{3}}{2}$; BC = $h \cdot \frac{1}{2}$ |
| 60 – 30 – 90<br>A – B – C | Hypotenuse $h$ (AB); AC = $h \cdot \frac{1}{2}$; BC = $h \cdot \frac{\sqrt{3}}{2}$ |

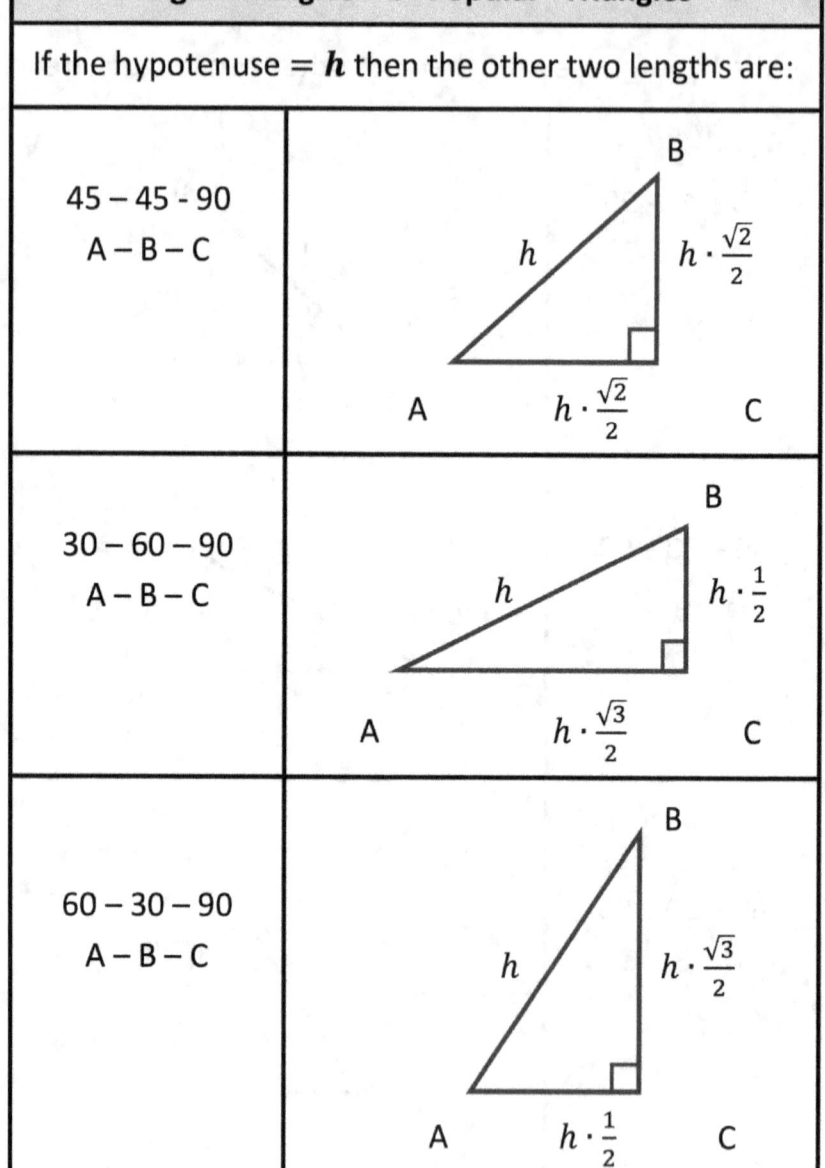

## 3 "Popular" Right Triangles, Hypotenuse = 1
## Confirm – Ex. 1

Use the Pythagorean Theorem to confirm the lengths.

### 45 – 45 - 90

$1^2 = \left(\frac{\sqrt{2}}{2}\right)^2 + \left(\frac{\sqrt{2}}{2}\right)^2$

$1 = \frac{2}{4} + \frac{2}{4}$   TRUE

### 30 – 60 – 90

$1^2 = \left(\frac{1}{2}\right)^2 + \left(\frac{\sqrt{3}}{2}\right)^2$

$1 = \frac{1}{4} + \frac{3}{4}$   TRUE

### 60 – 30 – 90

$1^2 = \left(\frac{\sqrt{3}}{2}\right)^2 + \left(\frac{1}{2}\right)^2$

$1 = \frac{3}{4} + \frac{1}{4}$

$1 = 1$   TRUE

### Right Triangles
### Use Pythagorean Theorem – Ex. 2

Given a right triangle with leg lengths of 3 and 4
Use the Pythagorean Theorem to find the hypotenuse.

| Make a sketch | 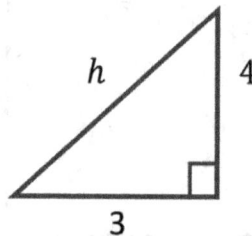 |
|---|---|
| Use Pythagorean Theorem $$c^2 = a^2 + b^2$$ | $h^2 = 3^2 + 4^2$ <br> $h^2 = 9 + 16$ <br> $h^2 = 25$ <br> $h = \pm\sqrt{25}$ <br> $h = 5$      Answer |

Note #1: Disregard the negative solution.

Note #2: The 3-4-5 triangle is also very "Popular." It appears in many standardized exams. It may appear as a 3-4-5 triangle, or multiples of it (which are similar triangles). For example: 6-8-10 or 9-12-15 ...

# The Unit Circle

## Right Triangles in a Circle
## With Radius = 1

There are an infinite number of triangles inside any circle. The diagram below shows 3 "Popular" or standard triangles within a circle, with radius = 1.

The 3 popular triangles are:
    45-45-90,   30-60-90,   and   60-30-90

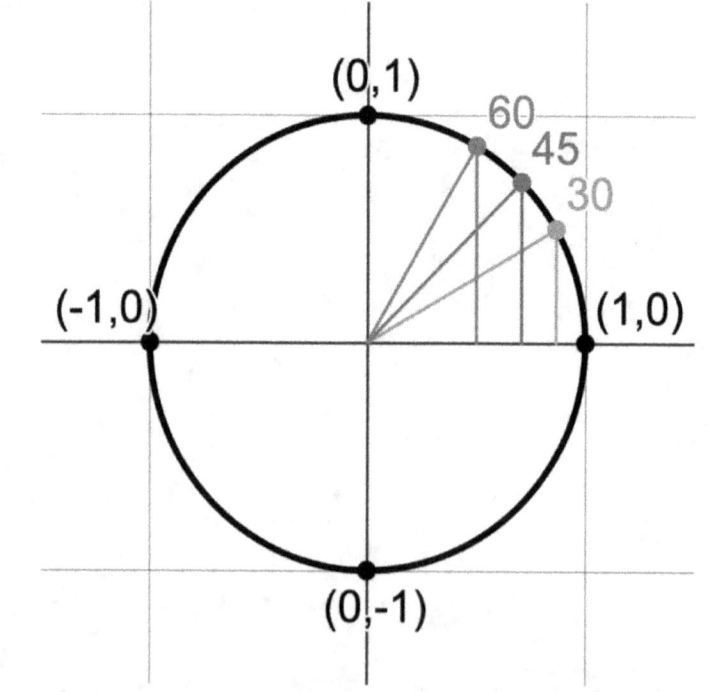

Can you find the coordinates where the 3 triangles intersect the circle? (See next page!)

## Right Triangles in a Circle
## With Radius = 1  (Showing Coordinates)

There are an infinite number of triangles inside any circle. The diagram below shows 3 standard triangles within a circle, with radius = 1.

The 3 popular triangles are:
    45-45-90,   30-60-90,   and   60-30-90

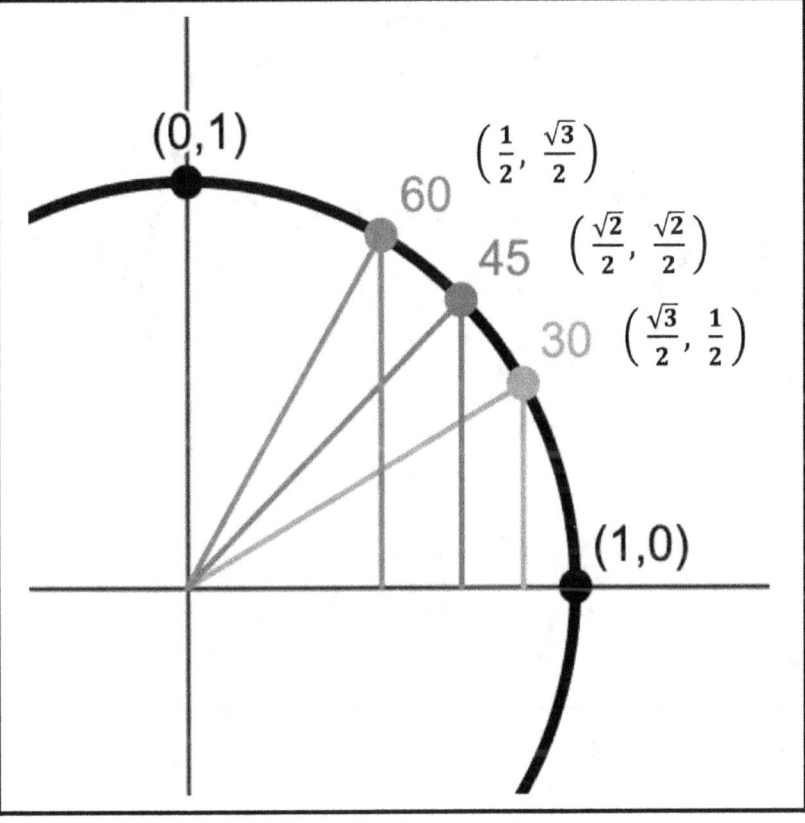

## The Unit Circle
## With Radius = 1   (Quadrant 1)

There are an infinite number of triangles inside any circle. The diagram below shows 3 standard triangles within a circle, with radius = 1.

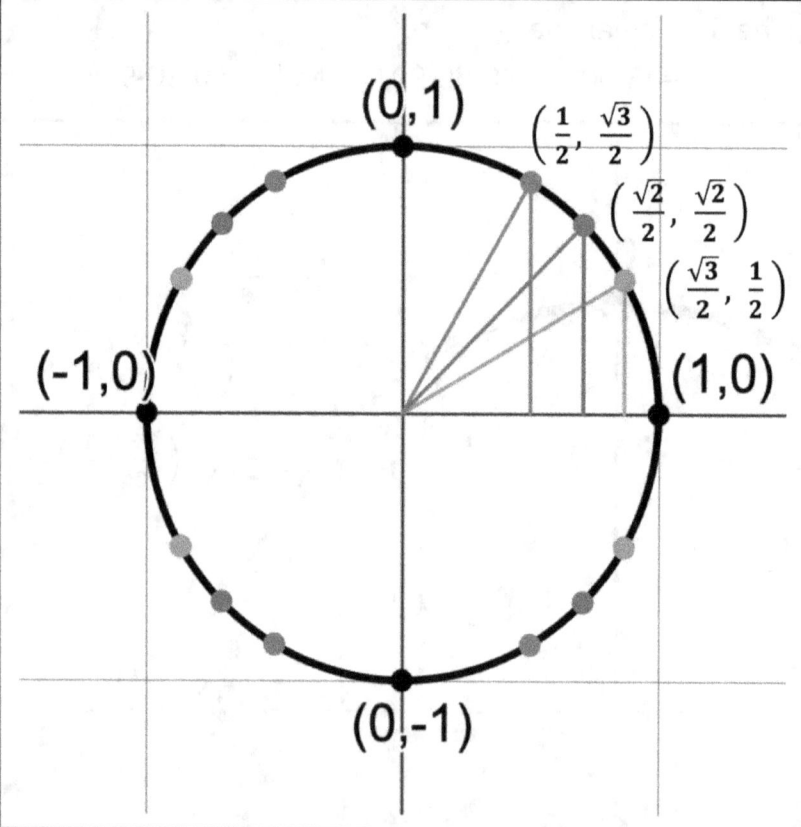

Can you find the coordinates of all points on the unit circle? (See next page.)

## The Unit Circle
## With Radius = 1

There are an infinite number of triangles inside any circle. The diagram below shows 3 right triangles (30°, 45°, 60°) within a circle, with radius = 1.

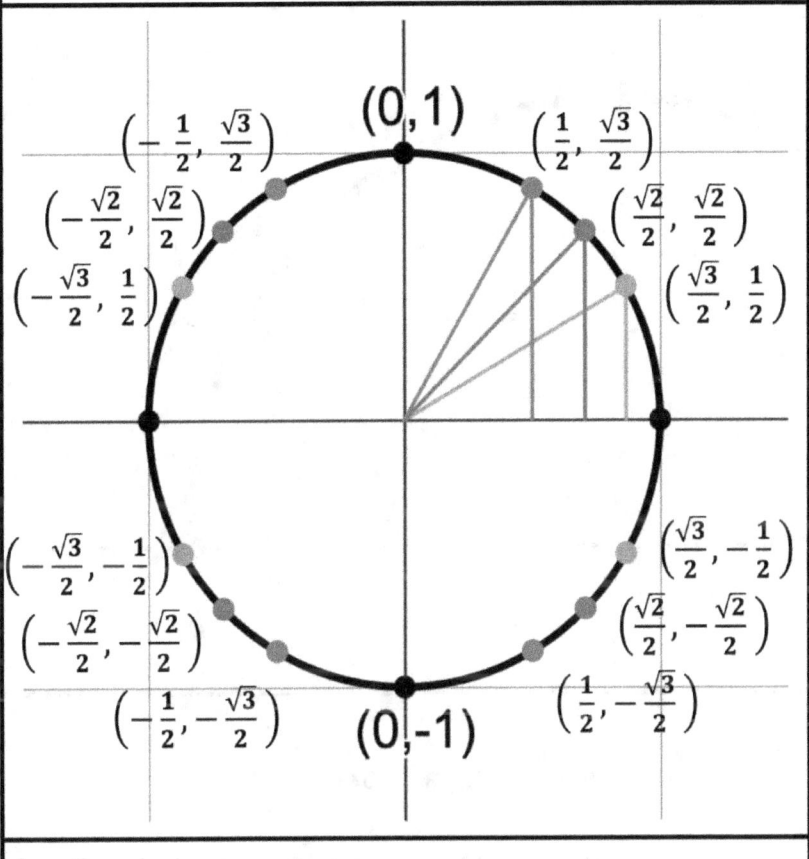

Notice the symmetry.

## The Unit Circle – Measuring Angles

Angles, within a circle, are measured from the *x* axis, as shown below.

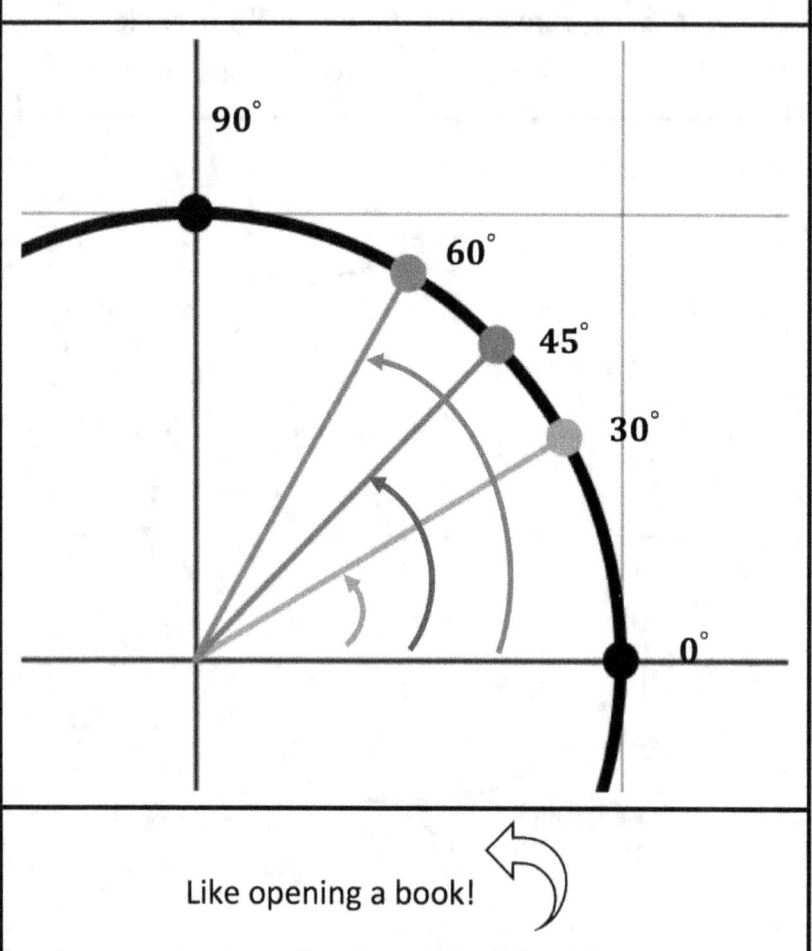

Like opening a book!

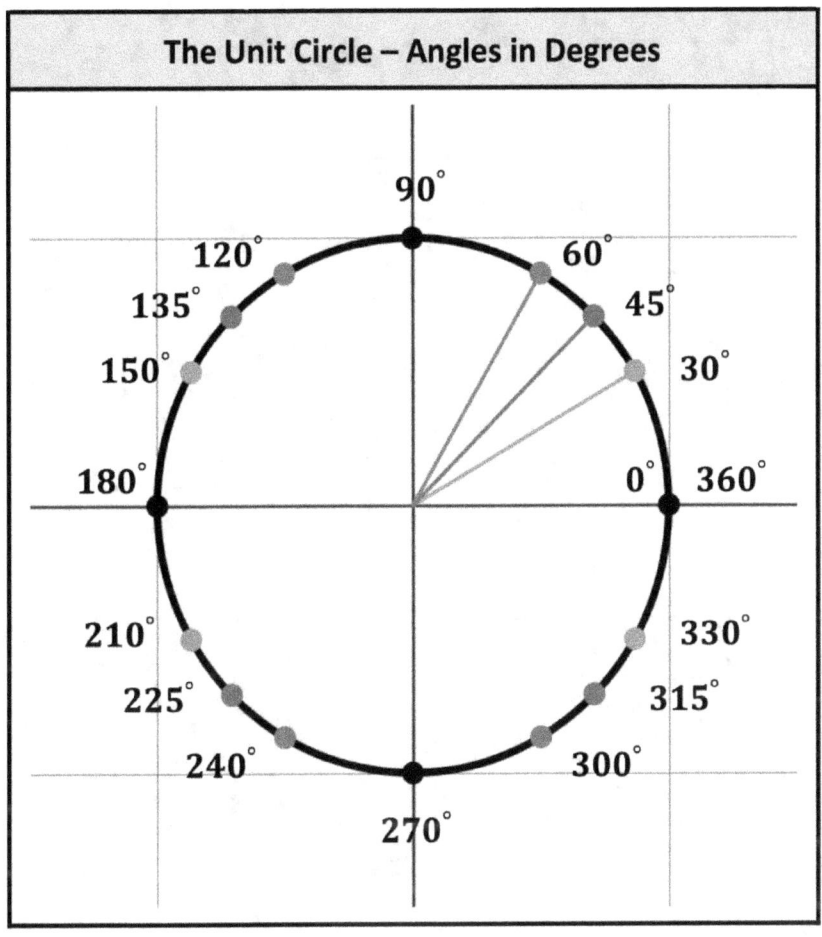

## The Unit Circle – Angles in Radians

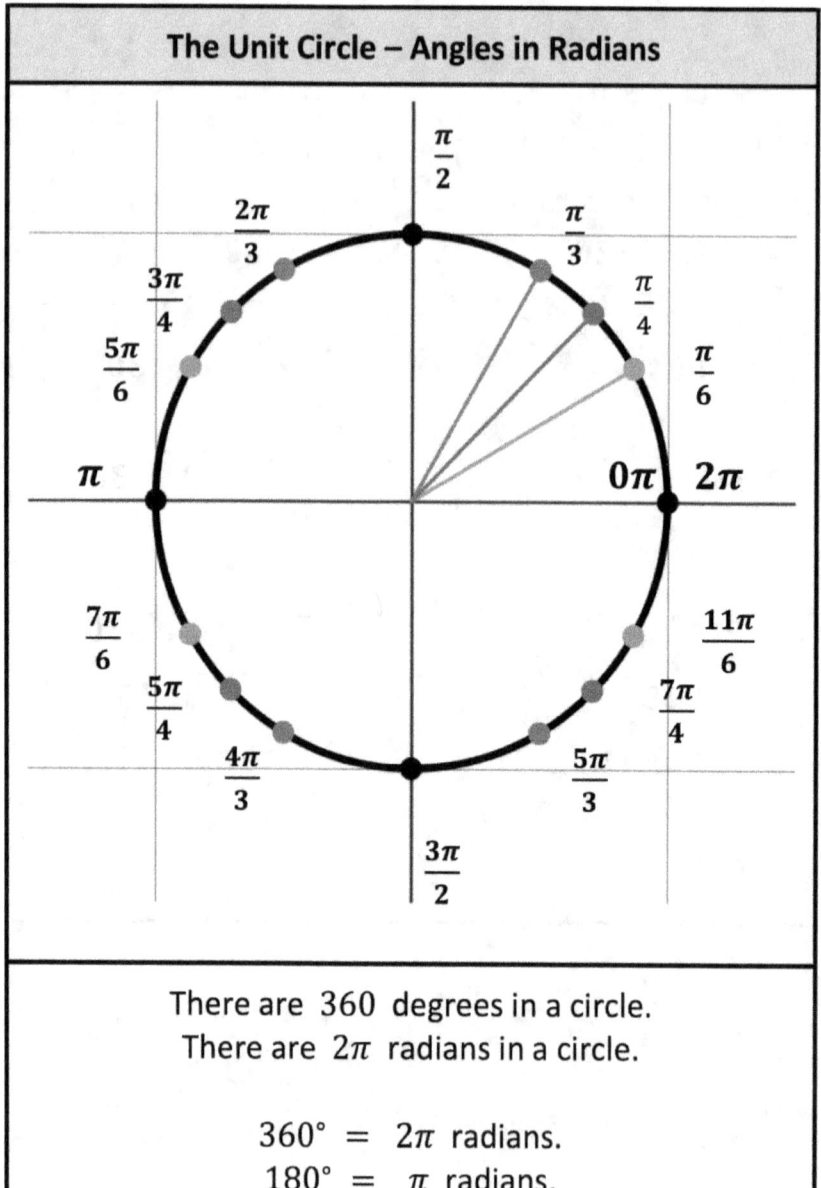

There are 360 degrees in a circle.
There are $2\pi$ radians in a circle.

$360° = 2\pi$ radians.
$180° = \pi$ radians.

## The Unit Circle

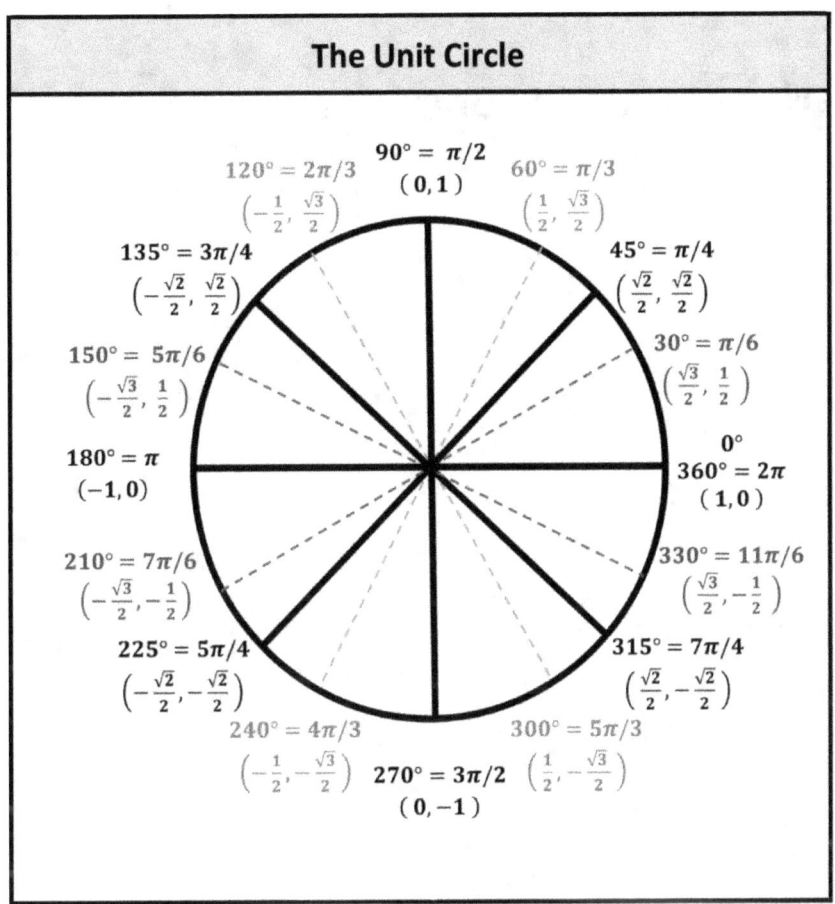

| Trig. Function | When $r = 1$ |
|---|---|
| $x = r \cos \theta$ | $x = \cos \theta$ |
| $y = r \sin \theta$ | $y = r \sin \theta$ |

# Trigonometric Functions

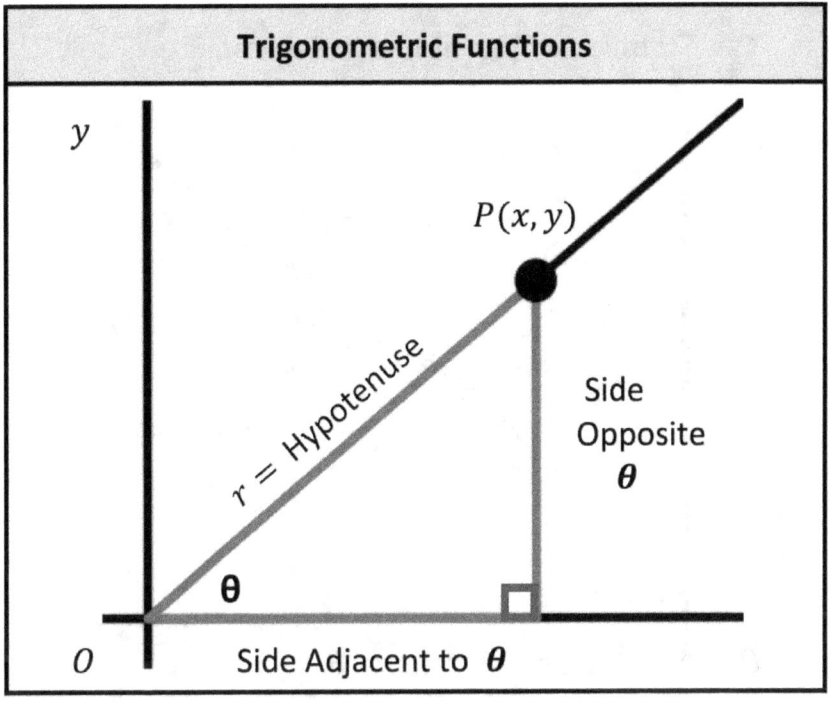

## 6 Trigonometric Functions

| | |
|---|---|
| $\cos\theta = \dfrac{Adjacent}{r} = \dfrac{x}{r}$ | $\sec\theta = \dfrac{1}{\cos\theta} = \dfrac{r}{x}$ |
| $\sin\theta = \dfrac{Opposite}{r} = \dfrac{y}{r}$ | $\csc\theta = \dfrac{1}{\sin\theta} = \dfrac{r}{y}$ |
| $\tan\theta = \dfrac{Opposite}{Adjacent} = \dfrac{y}{x}$ | $\cot\theta = \dfrac{1}{\tan\theta} = \dfrac{x}{y}$ |

| | |
|---|---|
| $x = r\cos\theta$ | $y = r\sin\theta$ |

## Inverse Trigonometric Functions

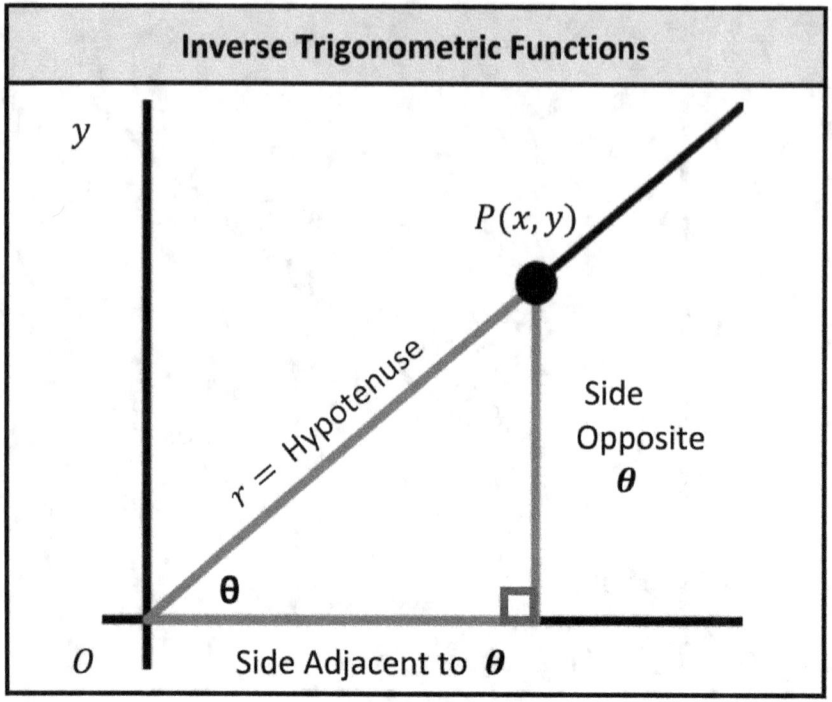

| Inverse Trig. Functions. (2 notations) |
|---|
| $\theta = \cos^{-1}\left(\dfrac{x}{r}\right) = arcCos\left(\dfrac{x}{r}\right)$ |
| $\theta = \sin^{-1}\left(\dfrac{y}{r}\right) = arcsin\left(\dfrac{y}{r}\right)$ |
| $\theta = \tan^{-1}\left(\dfrac{y}{x}\right) = arctan\left(\dfrac{y}{x}\right)$ |
| $\theta = \cos^{-1}\left(\dfrac{x}{r}\right)$ = "The angle whose cosine is $\left(\dfrac{x}{r}\right)$ " |

# Trigonometric Functions – Ex. 1

Given: An acute angle $\theta$ in standard position and its terminal side passes through $P(4,5)$.
Find: $\cos\theta, \sin\theta,$ and $\tan\theta$.

Notes:
- Acute means the angle is $< 90°$
- Obtuse means the angle is $> 90°$
- Standard position means the vertex is at the orgin.

| Make a sketch and use Pythagorean Theorm to find the Hypotenuse | | $r^2 = 4^2 + 5^2$ $r^2 = 16 + 25$ $r = \pm\sqrt{41}$ $r = \sqrt{41}$ |
|---|---|---|
| $\cos\theta$ | $\cos\theta = \frac{x}{r} = \frac{4}{\sqrt{41}}\left(\frac{\sqrt{41}}{\sqrt{41}}\right) = \frac{4\sqrt{41}}{41}$ | |
| $\sin\theta$ | $\sin\theta = \frac{y}{r} = \frac{5}{\sqrt{41}}\left(\frac{\sqrt{41}}{\sqrt{41}}\right) = \frac{5\sqrt{41}}{41}$ | |
| $\tan\theta$ | $\tan\theta = \frac{y}{x} = \frac{5}{4}$ | |

## Trigonometric Functions – Ex. 2

Given: An acute angle $\theta$ in standard position and its terminal side passes through $P(5,12)$.

Find: The six trigonometric functions.

| Make a sketch and use Pythagorean Theorem to find the Hypotenuse | 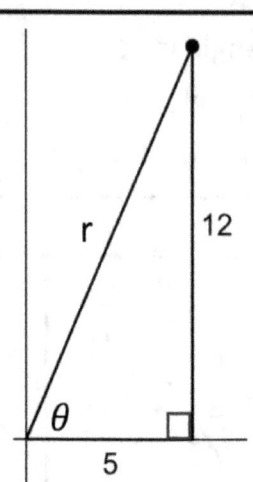 | $r^2 = 5^2 + 12^2$ <br> $r^2 = 25 + 144$ <br> $r = \pm\sqrt{169}$ <br> $r = 13$ |
|---|---|---|

### Six Trigonometric Functions

| | |
|---|---|
| $\cos\theta = \dfrac{x}{r} = \dfrac{5}{13}$ | $\sec\theta = \dfrac{1}{\cos\theta} = \dfrac{13}{5}$ |
| $\sin\theta = \dfrac{y}{r} = \dfrac{12}{13}$ | $\csc\theta = \dfrac{1}{\sin\theta} = \dfrac{13}{12}$ |
| $\tan\theta = \dfrac{y}{x} = \dfrac{12}{5}$ | $\cot\theta = \dfrac{1}{\tan\theta} = \dfrac{5}{12}$ |

## Trigonometric Functions – Ex. 3

Given: $\sin\theta = \frac{1}{3}$ and $\theta$ is an acute angle.

Find: $\cos\theta$ and $\tan\theta$

| | |
|---|---|
| If $\sin\theta = \frac{1}{3}$<br><br>Then $\frac{y}{r} = \frac{1}{3}$<br><br>So:<br>$y = 1$ and $r = 3$ | 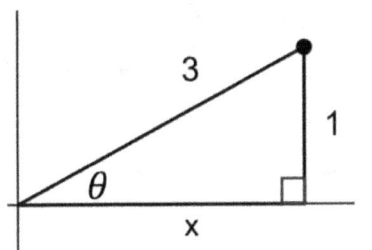 |
| Use Pythagorean Theorem to find the missing side. | $r^2 = x^2 + y^2$<br>$3^2 = x^2 + 1^2$<br>$x^2 = 3^2 - 1^2$<br>$x^2 = 9 - 1$<br>$x = \pm\sqrt{8}$<br>$x = 2\sqrt{2}$ |

$$\cos\theta = \frac{x}{r} = \frac{2\sqrt{2}}{3}$$

$$\tan\theta = \frac{y}{x} = \frac{1}{2\sqrt{2}}\left(\frac{\sqrt{2}}{\sqrt{2}}\right) = \frac{\sqrt{2}}{4}$$

## Trigonometric Functions – Ex. 4

Use information in the diagram to find the lengths of sides.

$\overline{BC}$ and $\overline{AB}$

Use a calculator

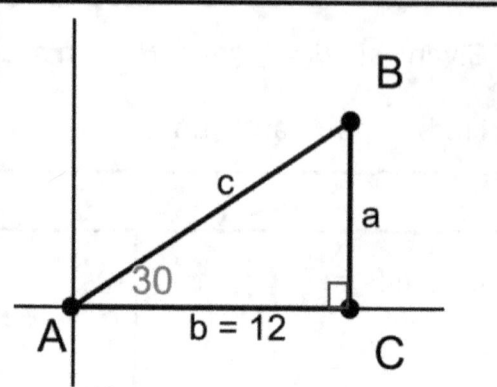

| Find $a$<br><br>Note: $a = \overline{BC}$ | $\tan 30 = \dfrac{a}{12}$<br><br>$a = 12 \cdot \tan 30$<br><br>$a = 12 \cdot \left(\dfrac{\sqrt{3}}{3}\right) = 4\sqrt{3}$ |
|---|---|
| Find $c$<br><br>Note: $c = \overline{AB}$ | $\cos 30 = \dfrac{b}{c}$<br><br>$c = \dfrac{b}{\cos 30}$<br><br>$c = \dfrac{12}{\left(\dfrac{\sqrt{3}}{2}\right)} = \dfrac{24}{\sqrt{3}}$<br><br>$c = \dfrac{24}{\sqrt{3}} \left(\dfrac{\sqrt{3}}{\sqrt{3}}\right) = \dfrac{24\sqrt{3}}{3} = 8\sqrt{3}$ |

## Trigonometric Functions – Ex. 5a

| Use information in the diagram to find the measure of angle $A$  Use a calculator | 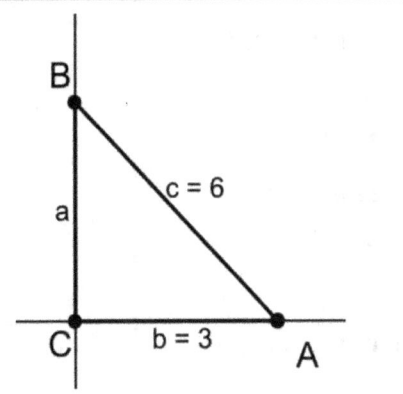 |
|---|---|
| **What do we know?**  Hint: Use given information. | For $\theta = A$  $\cos \theta = \dfrac{adjacent}{hypotenuse}$  $\cos A = \dfrac{b}{c}$  $\cos A = \dfrac{3}{6} = \dfrac{1}{2}$ |
| We know angle $A$ Is an angle whose cosine is $\dfrac{1}{2}$ | $A = \cos^{-1}\left(\dfrac{1}{2}\right)$  $A = 60°$   Use calculator.  Make sure calculator MODE is in Degrees. |

## Trigonometric Functions – Ex. 5b

| | |
|---|---|
| Use information in the diagram to find the measure of angle $B$<br><br>Use a calculator | 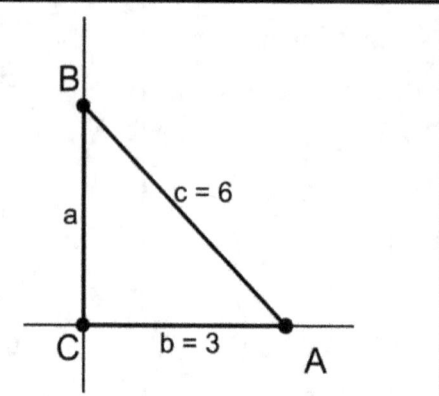 |

| What do we know?<br><br>Hint: Use given information. | For $\theta = B$<br><br>$\sin \theta = \dfrac{opposite}{hypotenuse}$<br><br>$\sin B = \dfrac{b}{c}$<br><br>$\sin B = \dfrac{3}{6} = \dfrac{1}{2}$ |
|---|---|
| We know angle $B$<br>Is an angle whose<br>sine is $\dfrac{1}{2}$ | $B = \sin^{-1}\left(\dfrac{1}{2}\right)$<br><br>$B = 30°$    Use calculator.<br><br>Make sure calculator MODE is in Degrees. |

# Trig Functions of General Angles

## Trigonometric Functions of General Angles

Place $\theta$ in standard position, choose a point $P(x, y)$ on the terminal side of $\theta$, and let $r =$ the distance $\overline{OP}$.

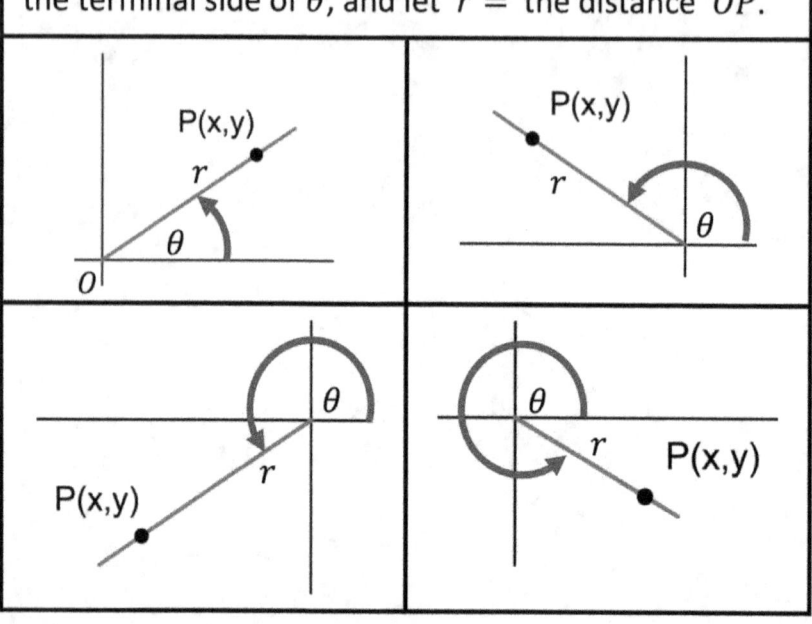

| | |
|---|---|
| $\cos\theta = \dfrac{x}{r}$ | $\sec\theta = \dfrac{1}{\cos\theta} = \dfrac{r}{x}$ ; $x \neq 0$ |
| $\sin\theta = \dfrac{y}{r}$ | $\csc\theta = \dfrac{1}{\sin\theta} = \dfrac{r}{y}$ ; $y \neq 0$ |
| $\tan\theta = \dfrac{y}{x}$ ; $x \neq 0$ | $\cot\theta = \dfrac{1}{\tan\theta} = \dfrac{x}{y}$ ; $y \neq 0$ |

$$r = \sqrt{x^2 + y^2}$$

## Trigonometric Functions of General Angles
### Reference Angle

When $\theta$ is in standard position, we call $\alpha$ the reference angle. Where $\alpha$ is an acute angle that represents $\theta$. See diagrams, below.

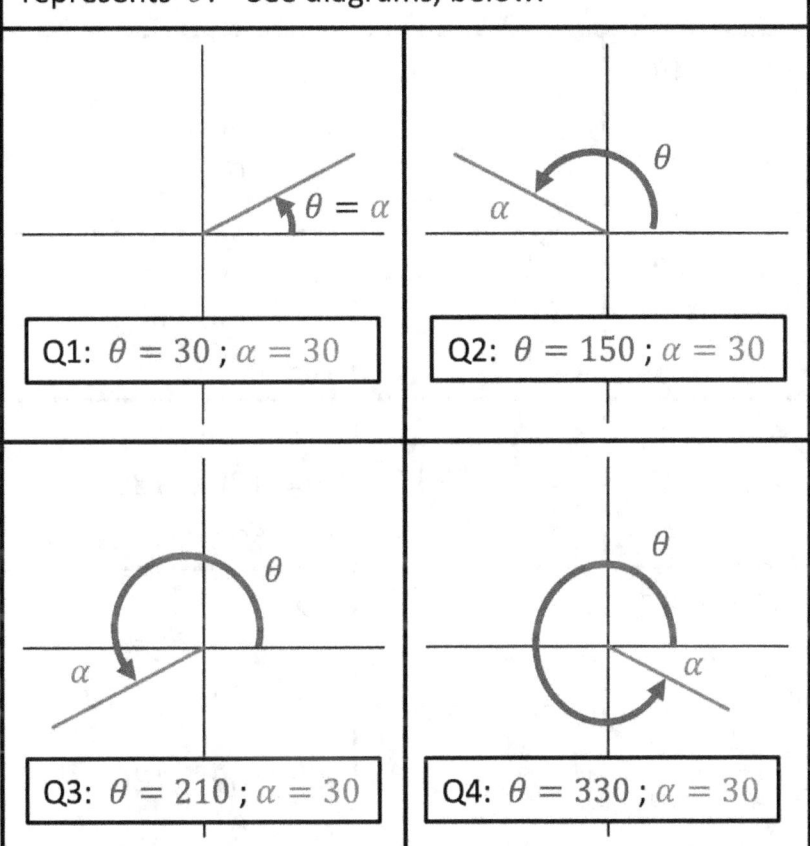

Q1: $\theta = 30$ ; $\alpha = 30$

Q2: $\theta = 150$ ; $\alpha = 30$

Q3: $\theta = 210$ ; $\alpha = 30$

Q4: $\theta = 330$ ; $\alpha = 30$

## Trigonometric Functions of General Angles
### Reference Angles – Ex. 1

Find the quadrant and measure of the reference angles, $\alpha$, for the following angles:

| $\theta = 140°$ | | Q2<br>$\alpha = 180 - 140$<br>$\alpha = 40°$ |
|---|---|---|
| $\theta = 300°$ | | Q4<br>$\alpha = 360 - 300$<br>$\alpha = 60°$ |
| $\theta = -135°$ | | Q3<br>$\alpha = 180 - 135$<br>$\alpha = 45°$ |
| $\theta = 480°$ | | Q2<br>$\theta = 480 - 360$<br>$\theta = 120$<br><br>$\alpha = 180 - 120$<br>$\alpha = 60°$ |

## Trigonometric Functions of General Angles
## Reference Angles – Ex. 2

Find the quadrant and exact value of the following:
$\tan 330°$ and $\csc(-225°)$

| $\theta = 330°$ | | Q4 <br> $\alpha = 360 - 330$ <br> $\alpha = 30°$ |
|---|---|---|
| | $\tan 30° = \dfrac{\sqrt{3}}{3} = \dfrac{y}{x}$ <br> In Q4, $(x, y) = (+, -)$ <br> $\tan 330° = -\dfrac{\sqrt{3}}{3}$      Answer | |

| $\theta = -225°$ | | Q2 <br> $\alpha = 225 - 180$ <br> $\alpha = 45°$ |
|---|---|---|
| | $\csc 45 = \dfrac{1}{\sin 45} = \dfrac{2}{\sqrt{2}}\left(\dfrac{\sqrt{2}}{\sqrt{2}}\right) = \sqrt{2} = \dfrac{r}{y}$ <br> In Q2, $(x, y) = (-, +)$ <br> $\csc(-225°) = \sqrt{2}$      Answer | |

## Trigonometric Functions of General Angles
## Reference Angles – Ex. 3

Find the five other trigonometric functions of $\theta$ if ...

$$\cos\theta = -\frac{2}{5} \quad \text{and} \quad 180° < \theta < 360°$$

---

Negative cosine tells us $\theta$ is in Q2 or Q3.

$180° < \theta < 360°$ tells us $\theta$ is in Q3 or Q4.

Therefore, $\theta$ is in Q3.    In Q3, $(x, y) = (-, -)$

| Sketch a reference angle in Q1, based on $\cos\alpha = \frac{2}{5} = \frac{x}{r}$ | $x^2 + y^2 = r^2$ $2^2 + y^2 = 5^2$ $y = \pm\sqrt{5^2 - 2^2}$ $y = \sqrt{21}$ | 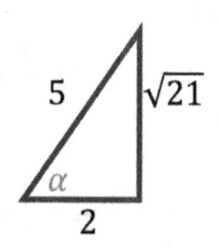 |

---

Remember:  In Q3, $(x, y) = (-, -)$

| $\cos\theta = \frac{x}{r} = -\frac{2}{5}$ | $\sec\theta = \frac{1}{\cos\theta} = -\frac{5}{2}$ |
|---|---|
| $\sin\theta = \frac{y}{r} = -\frac{\sqrt{21}}{5}$ | $\csc\theta = \frac{1}{\sin\theta} = -\frac{5}{\sqrt{21}} = -\frac{5\sqrt{21}}{21}$ |
| $\tan\theta = \frac{y}{x} = \frac{\sqrt{21}}{2}$ | $\cot\theta = \frac{1}{\tan\theta} = \frac{2}{\sqrt{21}} = \frac{2\sqrt{21}}{21}$ |

## Trigonometric Functions of General Angles – Ex. 4

A sailor sees a light tower, with the top of the tower at an angle of elevation of 25.6°. After his ship has moved 1050 $m$ closer, the angle of elevation is 31.2°. What is the height of the light tower, above sea level?

| | |
|---|---|
| Make a sketch | 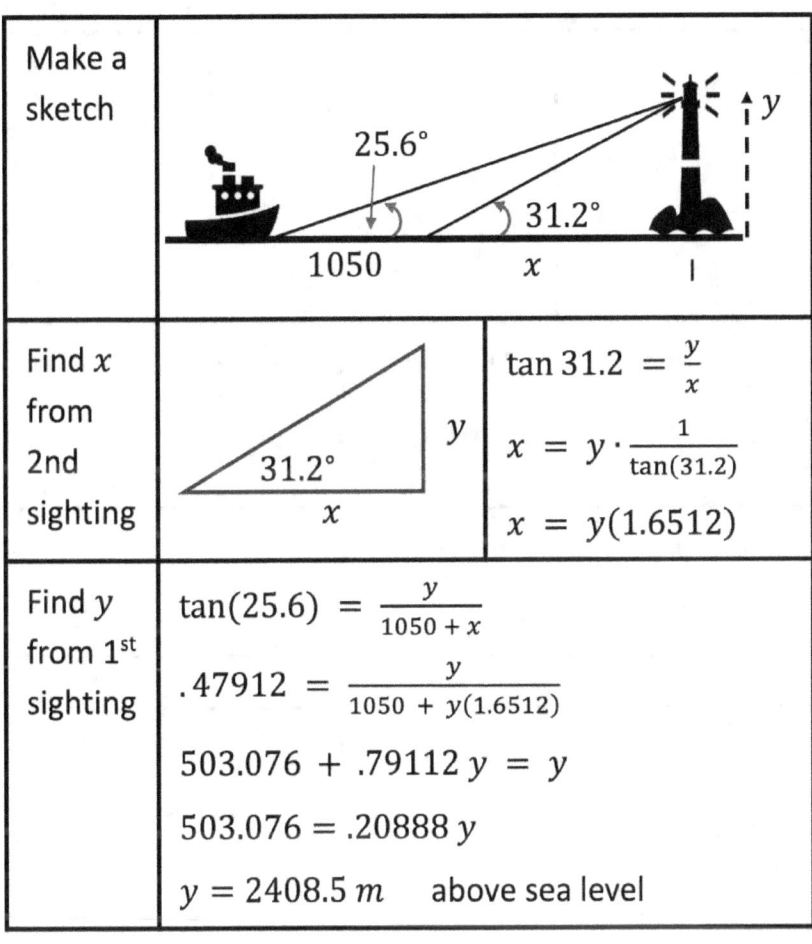 |
| Find $x$ from 2nd sighting | $\tan 31.2 = \dfrac{y}{x}$ <br><br> $x = y \cdot \dfrac{1}{\tan(31.2)}$ <br><br> $x = y(1.6512)$ |
| Find $y$ from 1st sighting | $\tan(25.6) = \dfrac{y}{1050 + x}$ <br><br> $.47912 = \dfrac{y}{1050 + y(1.6512)}$ <br><br> $503.076 + .79112\, y = y$ <br><br> $503.076 = .20888\, y$ <br><br> $y = 2408.5\ m$   above sea level |

## Trig Functions and Baseball – Ex. 5a

A baseball is hit at a point 3 feet above the ground at a velocity of 100 feet per second at an angle of 50°. Write equations to model the vertical height ($y$) and the horizontal distance ($x$) of the ball as a function of time.   Use Gravity = $32 \frac{ft}{s^2}$

| Make a sketch | 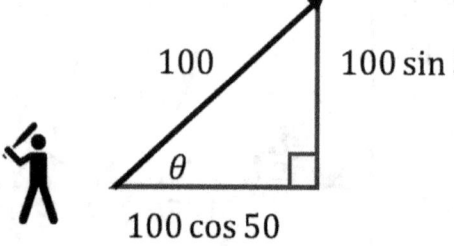 |
|---|---|

$y = -$ (gravity effect) + (vert. hit effect) + (init. height)

$y = -\left(\frac{32}{2}\right) t^2 + (100 \sin 50) t + (3)$

$x =$ (horizontal hit effect)

$x = (100 \cos 50) t$

| Note the units | $(vel.)(time) = \left(\frac{ft}{s}\right) \cdot (s) = ft$ |
|---|---|
| | $(accel.)(time^2) = \left(\frac{ft}{s^2}\right) \cdot (s^2) = ft$ |

## Trig Functions and Baseball – Ex. 5b

A baseball is hit at a point 3 feet above the ground at a velocity of 100 feet per second at an angle of $50°$.

There is a 5 foot fence, located 300 feet from home plate. Will the baseball clear the fence?

| Previously found | $y = -(16)t^2 + (100 \sin 50)t + (3)$ $x = (100 \cos 50)\,t$ |
|---|---|
| Find time when ball reaches fence | $300 = (100 \cos 50)\,t$ $t = \dfrac{300}{100 \cos 50} = 4.7$ seconds |
| Find height when ball reaches fence | $y = -(16)t^2 + (100 \sin 50)t + (3)$ $y = -(16)(4.7)^2 +$ $\quad + (100 \sin 50)(4.7) + (3)$ $y = -353.4 + 360.0 + 3$ $y = 9.6$ ft. |
| Answer | Yes! The baseball will clear the 5 ft. fence by 4.6 feet. |

# Law of Cosines

## Law of Cosines – For All Triangles

If you know the lengths of two sides and the measure of the included angle, then the **Law of Cosines** can be used to find the length of the other side.  **SAS**

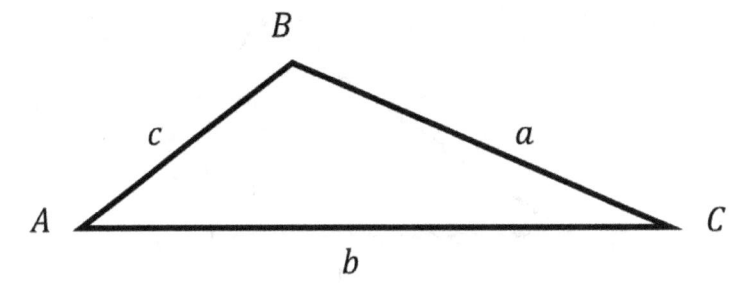

Note: The naming convention. Angles are marked with capitol letters. The sides, opposite from the angles are marked with lower-case letters.

The Law of Cosines is similar to the Pythagorean Theorem (for right triangles only). The Law of Cosines is for all triangles. The general form is:

$$a^2 = b^2 + c^2 - correction$$

$$a^2 = b^2 + c^2 - 2bc \cdot \cos A$$

$$b^2 = a^2 + c^2 - 2ac \cdot \cos B$$

$$c^2 = a^2 + b^2 - 2ab \cdot \cos C$$

## Law of Cosines – For All Triangles – Ex. 1

In $\triangle ABC$, $a = 10$, $b = 13$, $\angle C = 70°$
Find $c$ to four significant digits.

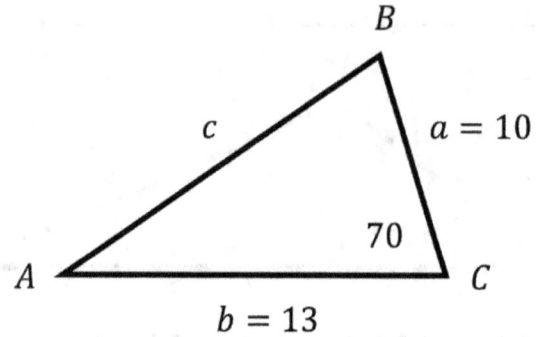

Note: We are looking for "c" so start with a version of the Law of Cosines with "C" on the ends (book-ends).

$c^2 = a^2 + b^2 - 2ab \cdot \cos C$

$c^2 = 10^2 + 13^2 - 2(10)(13) \cdot \cos(70)$

$c^2 = 100 + 169 - 260 \cdot \cos(70)$

$c^2 = 269 - 260 \cdot (.34202)$

$c^2 = 269 - 88.9252$

$c = \pm\sqrt{180.075} = 13.4192$

$c = 13.42$ to four significant digits.

## Law of Cosines – All Sides Known – Ex. 2

A triangular-shaped lot has three sides with lengths: 150, 120, $and$ 50 $meters$. Find the largest angle.

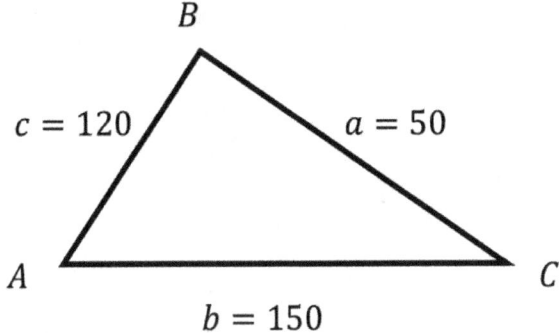

In our diagram, $b$ is the longest side. So, angle $B$ will be the largest angle. We are looking for "B" so start with a version of the Law of Cosines with "B" on the ends (book-ends).

$b^2 = a^2 + c^2 - 2ac \cdot \cos B$

$150^2 = 50^2 + 120^2 - 2(50)(120) \cdot \cos B$

$22500 = 16900 - 12000 \cdot \cos B$

$\cos B = \dfrac{22500 - 16900}{-12000} = \dfrac{5600}{-12000} = -.46667$

$B = \cos^{-1}(-.46667) = 117.82°$

# Law of Sines

## Law of Sines – For All Triangles

If you know the lengths of two sides and the measure of one angle, **NOT** between the two known sides, then the **Law of Sines** can be used to find the length of the other side.  **SSA (or ASS)**

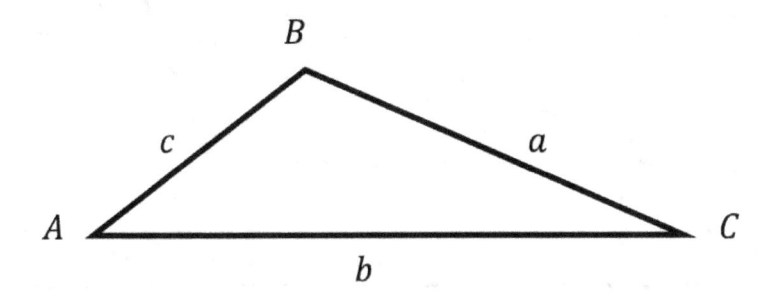

Note: The Law of Sines says the angles and the opposite sides are proportional.

$$\frac{a}{\sin A} = \frac{b}{\sin B} = \frac{c}{\sin C}$$

$$\frac{\sin A}{a} = \frac{\sin B}{b} = \frac{\sin C}{c}$$

### Law of Sines – For All Triangles – Ex. 1a

In $\triangle ABC$  $\angle A = 40°$ and $a = 15$
Find $\angle B$ if $b = 20$

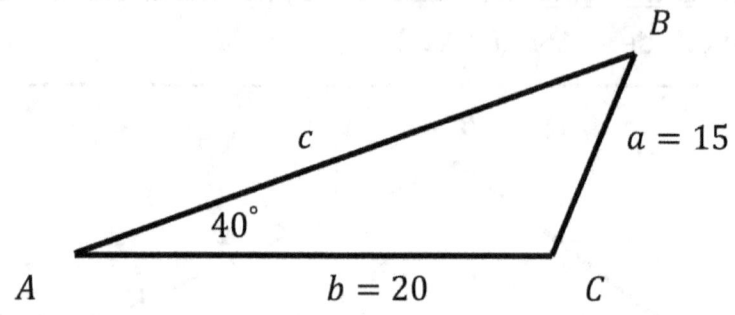

Note: We are looking for $B$ so use a version of the Law of Sines with $\sin B$ in numerator of first term. Also, since we know $\sin A$ and $a$, use that ratio.

$$\frac{\sin B}{b} = \frac{\sin A}{a}$$

$$\frac{\sin B}{20} = \frac{\sin(40)}{15}$$

$$\sin B = 20 \left(\frac{\sin(40)}{15}\right) = \frac{4}{3} \sin(40) = .85705$$

$$B = \sin^{-1}(.85705) = 59°$$

### Law of Sines – For All Triangles – Ex. 1b

In $\triangle ABC$ $\angle A = 40°$ and $a = 15$
Find $\angle B$ if $b = 20$

Previously, we found
$\sin B = .85705$

But, there are two angles
that have that $sin$

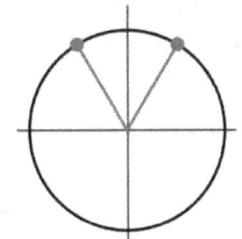

The calculator will return one value for
$B = \sin^{-1}(.85705) = 59°$
The default domain for the sin functions Is Q1 & Q4

There is another angle, in Q2, with the same sin
In Q2, the other angle is: $180 - 59 = 121°$
Therefore: $B = 59°$ or $121°$

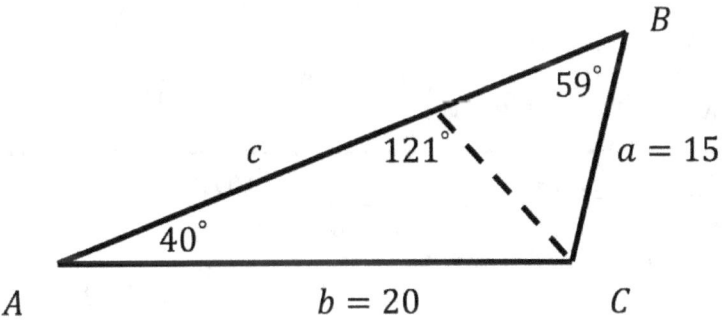

## Law of Sines – For All Triangles – Ex. 2

A 123 ft support wire for a large flag pole makes an angle for 61° with the ground. This wire will be replaced with a new wire with an angle of 46° What will be the length of the new wire?

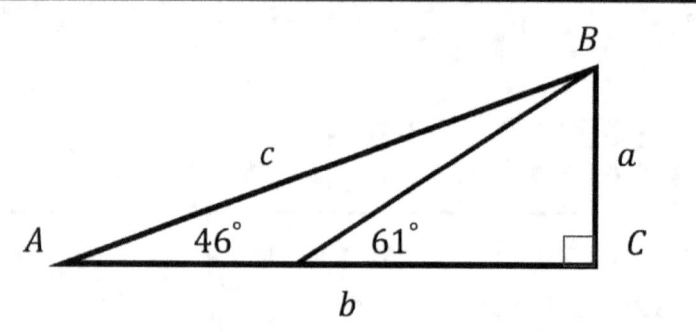

| Use this eqn. For both cases. | $\dfrac{c}{\sin C} = \dfrac{a}{\sin A}$ |
|---|---|
| Case 1: $A = 61°$  We know $c = 123$  Find $a$ (pole height) | $\dfrac{123}{\sin 90} = \dfrac{a}{\sin 61}$  $a = \sin 61 \left(\dfrac{123}{\sin 90}\right) = 107.6$ |
| Case 2: $A = 46°$  We know $a = 107.6$ | $\dfrac{c}{\sin 90} = \dfrac{107.6}{\sin 46}$  $c = (1)\left(\dfrac{107.6}{\sin 46}\right) = 149.6 \ ft.$ |

# General Triangles

## General Triangles

In this section, examples will be provided to demonstrate working with General Triangles when given the following information:

| Case | Given | Notes |
|---|---|---|
| SSS | Side-Side-Side | Fixed |
| SAS | Side-Angle-Side | Fixed |
| ASA, AAS | 2 Angles, 1 Side | Fixed |
| SSA | Side-Side-Angle | Ambiguous |

For all "Fixed" triangles, the approach is simple. If you are given a side and opposite angle, then use the "Law of Sines" to find additional information. The "Law of Sines" is a simpler equation to use than the "Law of Cosines."

For "Ambiguous" (or indeterminate) trinagles, be aware there are multiple possibilities. Visually place a hindge between the two given sides. There may be multiple possibilities for the angle between those two sides.

## Default Domains for Trigonometric Functions

It is important to be aware of the default domains when using calculators with trigonometric functions. The default domains are listed below.

| Trig Function | Default Domain |
|---|---|
| $\cos \theta$ and $\cos^{-1}(n)$ | Q1 & Q2 |
| $\sin \theta$ and $\sin^{-1}(n)$ | Q1 & Q4 |
| $\tan \theta$ and $\tan^{-1}(n)$ | Q1 & Q4 |

When using inverse trig functions, there are usually 2 answers within 360° (infinite solutions for all angles). A calculator will return only one solution, within the default domain. Use your underatanding of trig functions to determine the additional solution.

## General Triangles (SSS) – Ex. 1

Solve $\triangle ABC$, Given: $a = 4$, $b = 6$, $c = 5$

| | |
|---|---|
| Make a sketch | *(triangle with vertices A, B, C; sides c = 5, a = 4, b = 6)* |
| Can we use Law of Sines? | No! No angle and side pair. ☹ |
| Use Law of Cosines to get more info. | $a^2 = b^2 + c^2 - 2bc \cdot \cos A$ <br><br> $\cos A = \dfrac{a^2 - b^2 - c^2}{-2bc}$ <br><br> $A = \cos^{-1}\left(\dfrac{a^2 - b^2 - c^2}{-2bc}\right) = 41.4°$ |
| Now, use Law of Sines for more info. | $\dfrac{\sin B}{6} = \dfrac{\sin(41.4)}{4}$ <br><br> $B = \sin^{-1}\left(6 \cdot \dfrac{\sin(41.1)}{4}\right) = 82.7°$ |
| The last angle. | $C = 180 - 41.4 - 82.7 = 55.9°$ |

| | **General Triangles (SAS) – Ex. 2** |
|---|---|
| Solve $\triangle ABC$ , Given: $a = 8$, $c = 7$, $\angle B = 31.8°$ | |
| Make a sketch | Triangle with vertex $B$ at top (angle 31.8), $c = 7$ on left side to vertex $A$, $a = 8$ on right side to vertex $C$, and side $b$ along the bottom from $A$ to $C$. |
| Can we use Law of Sines? | No! No angle and side pair. ☹ |
| Use Law of Cosines to get more info. | $b^2 = a^2 + c^2 - 2ac \cdot \cos B$ <br> $b^2 = 17.812$ <br> $b = \sqrt{17.812} = 4.22$ |
| Now, use Law of Sines for more info. | $\dfrac{\sin A}{8} = \dfrac{\sin(31.8)}{4.22}$ <br> $A = \sin^{-1}\left(8 \cdot \dfrac{\sin(31.8)}{4.22}\right) = 87.4°$ |
| The last angle. | $C = 180 - 31.8 - 87.4 = 60.8°$ |

## General Triangles (AAS) – Ex. 3

Solve $\triangle ABC$, Given: $a = 40$, $\angle A = 45°$, $\angle C = 55°$

| | |
|---|---|
| Make a sketch | *Triangle with vertex $B$ at top, vertex $A$ at bottom-left with angle $45°$, vertex $C$ at bottom-right with angle $55°$. Side $a = 40$ opposite $A$, side $b$ opposite $B$, side $c$ opposite $C$.* |
| Get 3rd angle | $B = 180 - 45 - 55 = 80°$ |
| Can we use Law of Sines? | Yes! We have angle and side pair. ☺ |
| Use Law of Sines to get more info. | $\dfrac{c}{\sin(55)} = \dfrac{40}{\sin(45)}$ <br><br> $c = \sin 55 \left(\dfrac{40}{\sin 45}\right) = 46.3$ |
| Use Law of Sines again. | $\dfrac{b}{\sin(80)} = \dfrac{40}{\sin(45)}$ <br><br> $b = \sin 80 \left(\dfrac{40}{\sin 45}\right) = 55.7$ |

## General Triangles (ASS) – Ex. 4a

Solve $\triangle ABC$, Given: $b = 22$, $c = 30$, $\angle B = 30°$

| Make a sketch | *(sketch: triangle with vertices A, B, C; $\angle B = 30°$, side $c = 30$ from A to B, side $b = 22$ from A to C, side $a$ from B to C)* |
|---|---|
| WARNING!! "ASS" | AMBIGUOUS CASE!!! The angle, between the two given sides (Angle A) can be several angles. Threat it like a hindge. |
| Consider multiple possiblities | *(sketch: triangle showing hinge at A with two possible positions for side b = 22 reaching to C)* |

| | |
|---|---|
| \multicolumn{2}{c}{**General Triangles (ASS) – Ex. 4b**} |

Solve $\triangle ABC$, Given: $b = 22$, $c = 30$, $\angle B = 30°$

| | |
|---|---|
| Consider multiple possiblities | 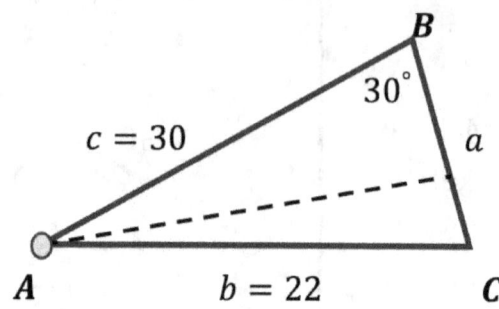 |
| Find $C$ Because we know $c = 30$ | $\dfrac{\sin C}{30} = \dfrac{\sin 30}{22}$<br><br>$\sin C = 30\left(\dfrac{\sin 30}{22}\right) = .681818$<br><br>$C = \sin^{-1}(.681818) = 43°$ |
| Find another value for angle $C$ | There are two angles that have $\sin = .681818$ 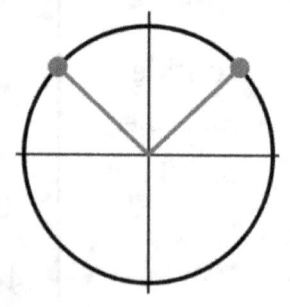<br><br>Also: $C = 180 - 43 = 137°$ |

## General Triangles (ASS) – Ex. 4c

Solve $\triangle ABC$, Given: $b = 22$, $c = 30$, $\angle B = 30°$

| Consider multiple possiblities | 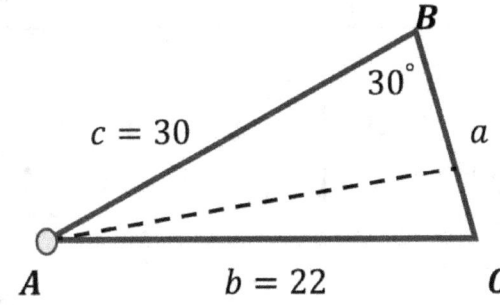 |
|---|---|
| Previously found | $C = 43°$    or    $C = 137°$ |

| Case #1<br><br>$C = 43°$ | $A = 180 - 30 - 43 = 107°$ |
|---|---|
| | $\dfrac{a}{\sin 107} = \dfrac{22}{\sin 30}$ |
| | $a = \sin 107 \left(\dfrac{22}{\sin 30}\right) = 42$ |

| Case #2<br><br>$C = 137°$ | $A = 180 - 30 - 137 = 13°$ |
|---|---|
| | $\dfrac{a}{\sin 13} = \dfrac{22}{\sin 30}$ |
| | $a = \sin 13 \left(\dfrac{22}{\sin 30}\right) = 9.9$ |

| | General Triangles (ASS) – Ex. 5a |
|---|---|
| | For: $\triangle ABC$, Given: $\angle A = 25°$, $c = 30$, $a = k$<br>Find values for $k$ so that $0, 1, 2$ triangles are possible. |
| Make a sketch | 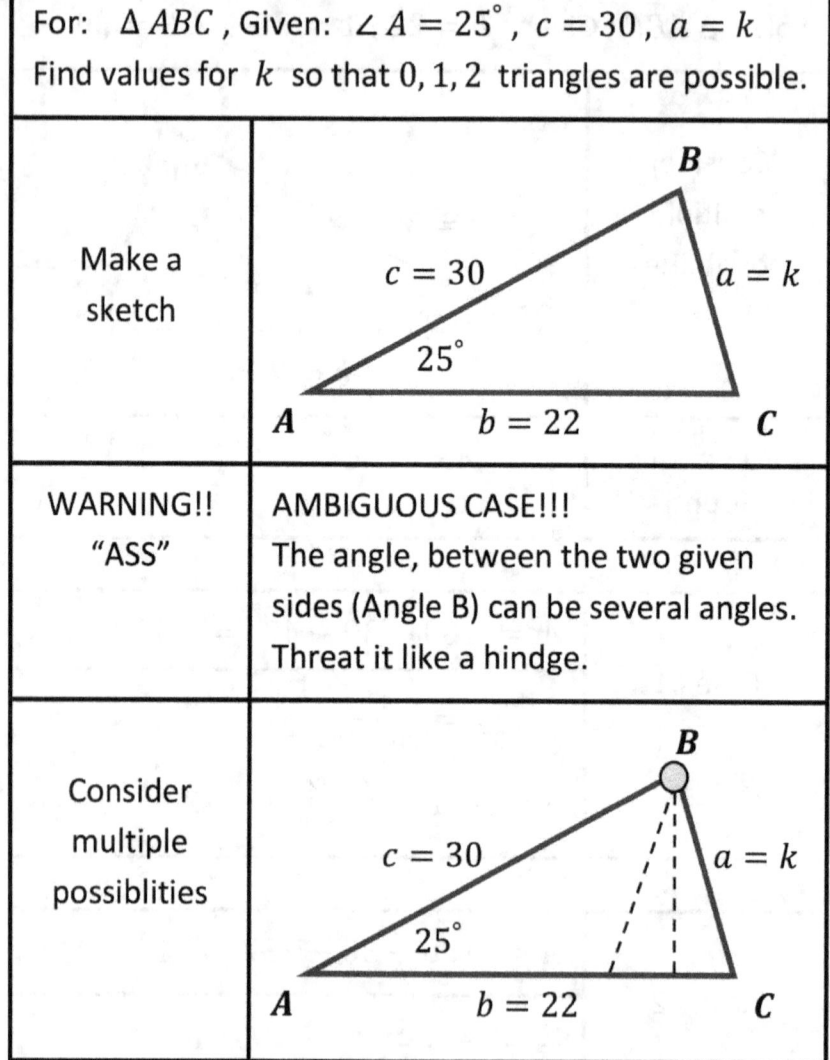 |
| WARNING!! "ASS" | AMBIGUOUS CASE!!!<br>The angle, between the two given sides (Angle B) can be several angles. Threat it like a hindge. |
| Consider multiple possiblities | |

## General Triangles (ASS) – Ex. 5b

For: $\triangle ABC$, Given: $\angle A = 25°$, $c = 30$, $a = k$
Find values for $k$ so that $0, 1, 2$ triangles are possible.

| | |
|---|---|
| Consider multiple possiblities | *(diagram: triangle with $c = 30$, $25°$ at $A$, $b = 22$, $a = k$)* |
| 0 Triangles Possible | If side $a$ is too short, then no triangle is possible. Side $a$ must be long enough to make a right angle at $C$. |
| | If $C = 90°$ (right triangle)<br>$a = r \cdot \sin \theta$<br>$a = 30 \cdot \sin 25 = 12.7$ |
| | 0 Triangles if: $k < 12.7$ |
| 2 Triangles Possible | If: $12.7 < a < 30$ then side $a$ could swing inward or outward to create two different triangles. |
| | 2 Triangles if: $12.7 < k < 30$ |

| General Triangles (ASS) – Ex. 5c ||
|---|---|
| For: $\triangle ABC$, Given: $\angle A = 25°$, $c = 30$, $a = k$ <br> Find values for $k$ so that 0, 1, 2 triangles are possible. ||
| Consider multiple possiblities | 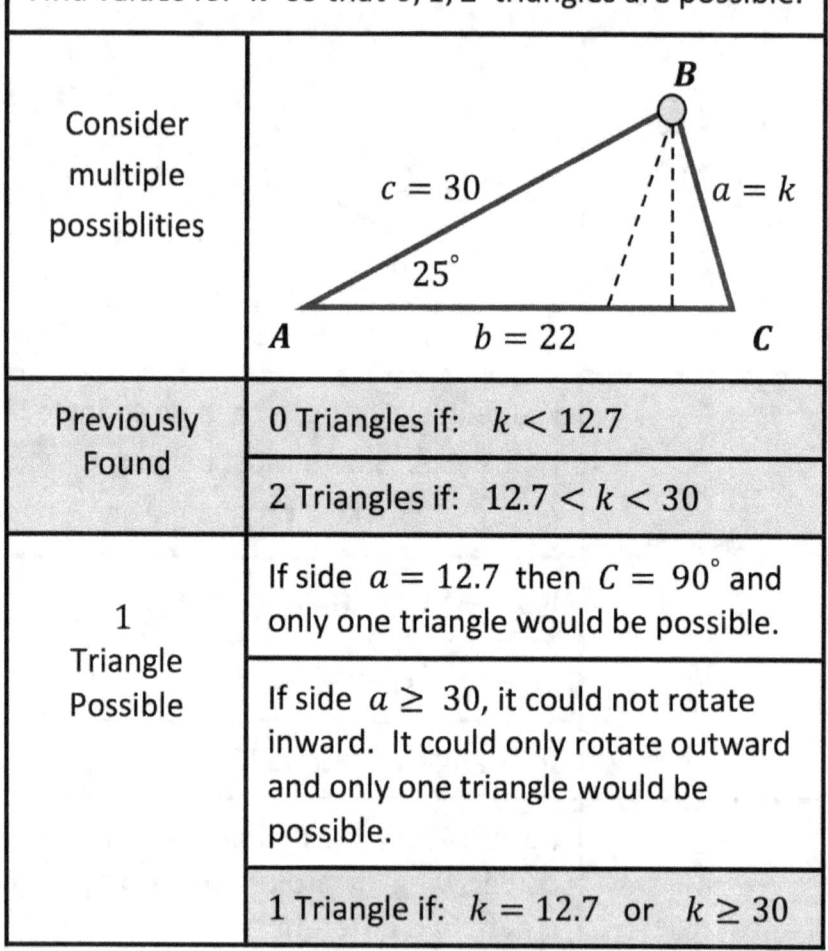 |
| Previously Found | 0 Triangles if: $k < 12.7$ |
|  | 2 Triangles if: $12.7 < k < 30$ |
| 1 Triangle Possible | If side $a = 12.7$ then $C = 90°$ and only one triangle would be possible. |
|  | If side $a \geq 30$, it could not rotate inward. It could only rotate outward and only one triangle would be possible. |
|  | 1 Triangle if: $k = 12.7$ or $k \geq 30$ |

# Bearings

## Bearings

In navigation, directions are usually given in terms of bearings. Some examples are given, below.

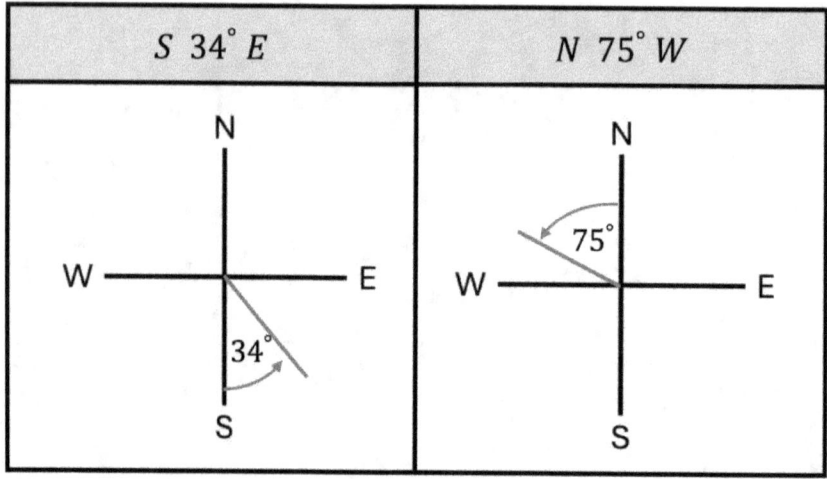

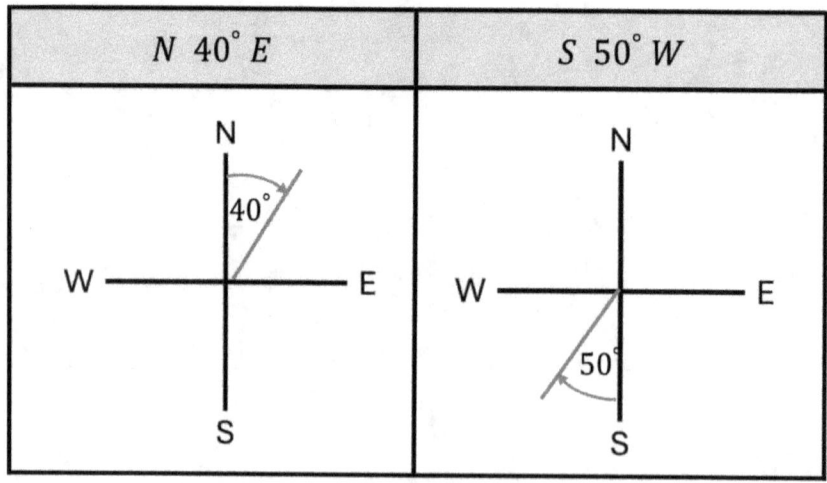

| | **Bearings – Ex. 1a** |
|---|---|

A ship leaves port at noon and heads due west at 20 knots (nautical mph). At 2 PM the ship changes course to $N\ 54°\ W$ as shown in the diagram. Find ship's bearing and distance from port at 3 PM.

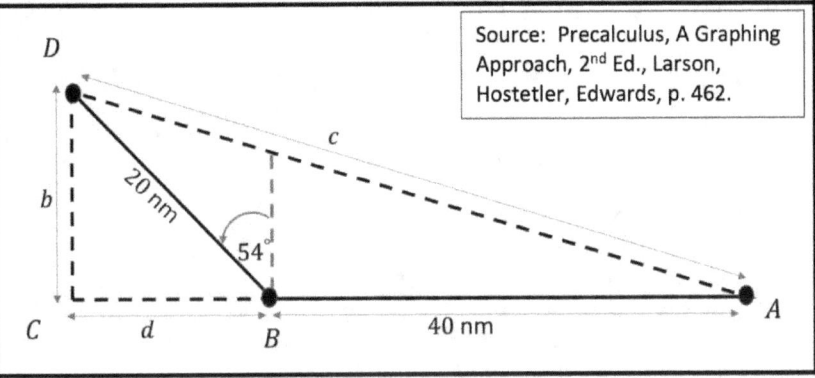

Source: Precalculus, A Graphing Approach, 2nd Ed., Larson, Hostetler, Edwards, p. 462.

| | |
|---|---|
| $\triangle BCD$ | $B = 90 - 54 = 36°$ <br> $b = 20 \sin 36 = 11.76$ <br> $d = 20 \cos 36 = 16.18$ |
| $\triangle ACD$ | $\tan A = \dfrac{b}{d+40} = \dfrac{11.76}{56.18} = .2093$ <br> $A = \tan^{-1}(.2093) = 11.82°$ <br> $D = 90 - 11.82 = 78.18°$ <br> $\dfrac{c}{\sin 90} = \dfrac{11.76}{\sin 11.82} \rightarrow c = 57.9$ nm |

## Bearings – Ex. 1b

A ship leaves port at noon and heads due west at 20 knots (nautical mph). At 2 PM the ship changes course to $N\ 54°\ W$ as shown in the diagram. Find ship's bearing and distance from port at 3 PM.

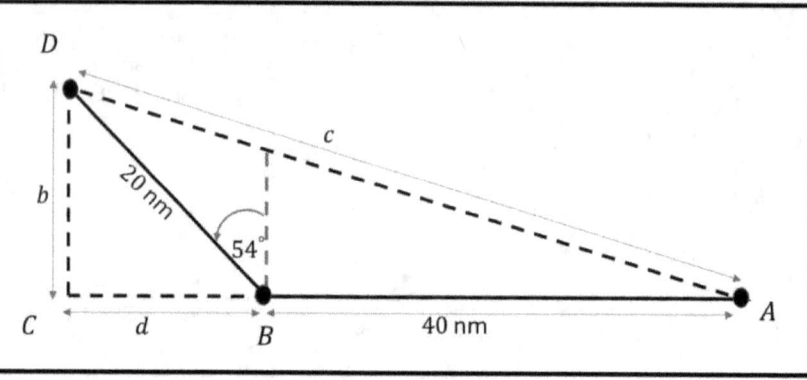

| Previously Found In $\triangle ACD$ | $A = 11.82°$ <br> $D = 78.18°$ | $c = 57.9$ nm |
|---|---|---|
| Distance from port | Distance $= 57.9$ nm | |
| Bearings from port (point A) | Angle with north-south line is $90° - 11.82° = 78.18°$ <br> Ship bearing is: $N\ 78.18°\ W$ | |

# References

## References

- Algebra and Trigonometry, Structure and Method, Book 2, Houghton Mifflin, Richard Brown, Mary Dolciani, Robert Sorgenfrey, Robert Kane, 1992.

- Mathematics, Structure and Method, Course 2, Mary Dolciani, Robert Sorgenfrey, John Grahm, McDougal Littell, 2001.

- Precalculus, A Graphing Approach, $2^{nd}$ Edition, Larson, Hostetler, Edwards, Houghton Mifflin Company, 1997.

- Essentials of College Algebra, Richard Aufmann, Richard Nation, Houghton Mifflin, 2006.

- One-Page Summaries for Algebra, Geometry, and Pre-Calculus, Kathryn Paulk, 2023.

- Complex Numbers and Polar Curves for Pre-Calc and Trig: With Problems and Detailed Solutions, Kathryn Paulk, 2023.

## Other Books by Kathryn Paulk

## Other Books by Kathryn Paulk

- Algebra 1 Help
- Algebra 2 Help
- Pre-Calculus and Trig Help
- College Algebra Help

- Calculus 1 Review in Bite-Size Pieces
- Calculus 2 Review in Bite-Size Pieces
- Calculus 3 Review in Bite-Size Pieces
- Differential Equations With Applications: Class Notes With Examples

- One-Page Summaries for Algebra, Geometry, and Pre-Calculus
- Graphing Functions Using Transformations for Algebra & Pre-Calc.
- Complex Numbers and Polar Curves
  For Pre-Calc and Trig:
  With Problems and Detailed Solutions
- Discrete and Continuous Probability Distributions: A Creative Comparison (V2)

- Teach Your Child to SWIM

## BIG MATH For Little Kids

## Workbooks for Young Children
## & Solution Manuals for Parents

- Introduction to Numbers

- Introduction to Fractions
  by Sharing Things

- Introduction to Counting and Fractions
  by Cooking Breakfast

- Learn About Fractions     *****
  by Baking Cookies

- Adding Big Numbers, Guessing Numbers
  and Secret Codes

- Learn to Graph by Riding Bikes
  on Graph Paper

www.ingramcontent.com/pod-product-compliance
Lightning Source LLC
Chambersburg PA
CBHW071826210526
45479CB00001B/5